Fuchsian Groups

Chicago Lectures in Mathematics Series
J. Peter May, Robert J. Zimmer, Spencer J. Bloch, and
Norman R. Lebovitz, editors

The Theory of Sheaves, by Richard G. Swan (1964)
Topics in Ring Theory, by I. N. Herstein (1969)
Fields and Rings, by Irving Kaplansky (1969; 2d ed. 1972)
Infinite Abelian Group Theory, by Philip A. Griffith (1970)
Topics in Operator Theory, by Richard Beals (1971)
Lie Algebras and Locally Compact Groups, by Irving Kaplansky (1971)
Several Complex Variables, by Raghavan Narasimhan (1971)
Torsion-Free Modules, by Eben Matlis (1973)
The Theory of Bernoulli Shifts, by Paul C. Shields (1973)
Stable Homotopy and Generalized Homology, by J. F. Adams (1974)
Commutative Rings, by Irving Kaplansky (1974)
Banach Algebras, by Richard Mosak (1975)
Rings with Involution, by I. N. Herstein (1976)
Theory of Unitary Group Representation, by George W. Mackey (1976)
Infinite-Dimensional Optimization and Convexity, by Ivar Ekeland and Thomas Turnbull (1963)
Commutative Semigroup Rings, by Robert Gilmer (1984)
Navier-Stokes Equations, by Peter Constantin and Ciprian Foias (1988)
Essential Results of Functional Analysis, by Robert J. Zimmer (1990)

Svetlana Katok

Fuchsian Groups

The University of Chicago Press
Chicago and London

SVETLANA KATOK is associate professor of mathematics at Pennsylvania State University.

The University of Chicago Press, Chicago 60637
The University of Chicago Press, Ltd., London

Printed in the United States of America

01 00 99 98 97 96 95 94 93 92 5 4 3 2 1

ISBN (cloth): 0-226-42582-7
ISBN (paper): 0-226-42583-5

Library of Congress Cataloging-in-Publication Data

Katok, Svetlana.
Fuchsian groups / Svetlana Katok.
p. cm. — (Chicago lectures in mathematics series)
Includes bibliographical references and index.
1. Fuchsian groups. I. Title. II. Series.
QA335.K38 1992
515′.93—dc20 92-6535
CIP

The paper used in this publication meets the minimum requirements of the American National Standard for Information Sciences—Permanence of Paper for Printed Library Materials, ANSI Z39.48-1984.

МОИМ РОДИТЕЛЯМ

To my parents Lucy and Boris

CONTENTS

PREFACE

This book originated from a one-quarter graduate course I taught at the California Institute of Technology and again at the University of California at Santa Cruz in 1989. The aim of this course was to present a self-contained theory of Fuchsian groups assuming no previous knowledge of the subject. Although the material in the first four chapters appears in many books, it either remains on an elementary level as a part of an introductory complex analysis course [JS], or is scattered in a presentation of a more sophisticated theory: discrete groups of Möbius transformations [B], discrete groups of isometries of spaces of constant curvature [VS], harmonic analysis on symmetric spaces [Te], quaternion algebras [Vi], or automorphic functions [GP, F, L, Sh]. My only contribution here was to present the material in as painless a way as possible, assuming only a basic knowledge of real and complex analysis and abstract algebra. In Chapter 5 I develop the theory of arithmetic Fuchsian groups: the modular group and its subgroups of finite index as well as Fuchsian groups derived from quaternion algebras. While that chapter is more specialized than the others, it is not written for specialists. Some knowledge of algebraic number theory would be helpful; however, I have tried to keep Chapter 5 as self-contained as possible, giving all the necessary definitions and presenting the theory of quaternion algebras on the level necessary for my purposes. Several examples appear throughout the book to illustrate important concepts and theorems. Each chapter contains a number of exercises; the hints for most of them are included at the end of the book.

Preface

Fuchsian groups–discrete groups of isometries of the hyperbolic plane–are a basic example of lattices in semisimple Lie groups. Their very concrete nature allows us to illustrate their many features that have far-reaching generalizations in geometry and number theory. The book therefore can be useful for graduate students specializing in a broad variety of areas of mathematics, including differential geometry, number theory, Lie theory, and representation theory, and can serve as an introduction to the more advanced works on those subjects. Capable graduate students should be able to read the book quite easily on their own.

I wish to express my appreciation to Peter Sarnak and Leonid Vaserstein for discussions on arithmetic groups. I would like to thank Ozlem Umamoglu and Yves Martin, graduate students from my class at the University of California at Santa Cruz, who read the manuscript and checked most of the excercises. I would also like to thank Jonathan Poritz, a graduate student at the University of Chicago, who read the first version of the manuscript with great care and made a number of useful comments that resulted in improvements in the exposition. The figures were created by my son Boris Katok, a recent graduate of the University of Chicago in political science. I greatly appreciate his help.

1. HYPERBOLIC GEOMETRY

1.1. The hyperbolic metric

Let $\mathbb{C}$ be the complex plane. We shall use the usual notations for the real and imaginary parts of $z=x+iy\in\mathbb{C}$, $\mathrm{Re}(z)=x$, $\mathrm{Im}(z)=y$. Our main object of study is the upper half-plane $\mathcal{H}=\{z\in\mathbb{C} \mid \mathrm{Im}(z)>0\}$. Equipped with the metric

$$ds=\frac{\sqrt{dx^2+dy^2}}{y} \tag{1.1.1}$$

it becomes a model of the *hyperbolic* or *Lobachevski* plane. We shall see that the role of *geodesics* (i.e. the shortest curves with respect to this metric) is played by straight lines and semicircles orthogonal to the real axis $\mathbb{R}=\{z\in\mathbb{C} \mid \mathrm{Im}(z)=0\}$. Any two points in $\mathcal{H}$ can be joined by a unique geodesic, and the distance between those points is measured along this geodesic. However, there is more than one geodesic passing through a given point z not in the geodesic L which does not intersect L (see Fig. 1); in fact, all geodesics through z outside of the shaded region do not intersect L. This means that this geometry in $\mathcal{H}$ is *non-Euclidean:* the fifth postulate of Euclid's *Elements*, the axiom of parallels, does not hold here. The metric (1.1.1) in $\mathcal{H}$ is called the *hyperbolic metric.* Let $I=[0, 1]$ and $\gamma\colon I\to\mathcal{H}$ be a piecewise differentiable path $\gamma=\{z(t)=x(t)+iy(t)\in\mathcal{H} \mid t\in I\}$. Then its *hyperbolic length* $h(\gamma)$ is given by

$$h(\gamma)=\int_0^1\frac{\sqrt{\left(\frac{dx}{dt}\right)^2+\left(\frac{dy}{dt}\right)^2}dt}{y(t)}=\int_0^1\frac{\left|\frac{dz}{dt}\right|dt}{y(t)}.$$

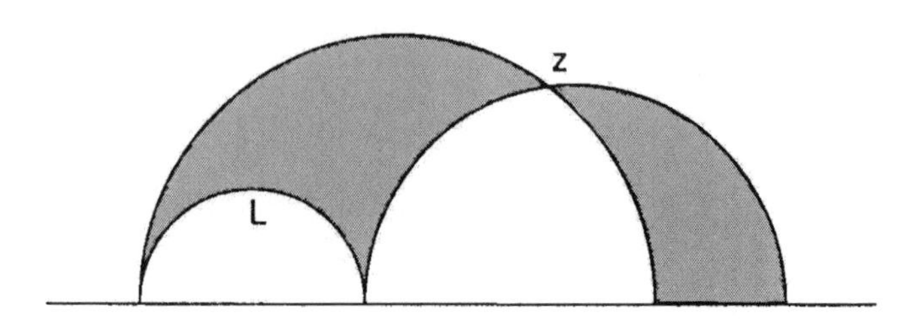

Fig. 1

DEFINITION. The *hyperbolic distance* $\rho(z, w)$ between two points $z, w \in \mathcal{H}$ is defined by the formula

$$\rho(z, w) = \inf h(\gamma),$$

where the infimum is taken over all γ joining z and w in $\mathcal{H}$.

It is easy to see that ρ is non-negative, is symmetric, and satisfies the triangle inequality

$$\rho(z,w) \leq \rho(z,\xi) + \rho(\xi,w),$$

i.e. is indeed a distance function on $\mathcal{H}$.

Let us consider a group of real matrices $g = \begin{bmatrix} a & b \\ c & d \end{bmatrix}$ with $\det(g) = ad - bc = 1$. As usual, $\operatorname{tr}(g) = a + d$ is the *trace* of a matrix g. This group is called the *unimodular group* and is denoted by $SL(2,\mathbb{R})$. The set of *fractional linear* (or *Möbius*) *transformations* of $\mathbb{C}$ onto itself of the form

$$\left\{ z \rightarrow \frac{az+b}{cz+d} \mid a,b,c,d \in \mathbb{R},\ ad-bc=1 \right\} \tag{1.1.2}$$

forms a group such that the product of two transformations corresponds to the product of corresponding matrices and the inverse corresponds to the inverse matrix. Each transformation T of the form (1.1.2) is represented by a pair of matrices $\pm g \in SL(2,\mathbb{R})$. Thus, the group of all transformations (1.1.2), called $PSL(2,\mathbb{R})$, is *isomorphic* to $SL(2,\mathbb{R})/\{\pm 1_2\}$, where 1_2 is the 2x2 identity matrix; we write $PSL(2,\mathbb{R}) \approx SL(2,\mathbb{R})/\{\pm 1_2\}$. The identity transformation in $PSL(2,\mathbb{R})$ will be denoted by Id. We have $tr(-g) = -tr(g)$, so that

$$tr^2(T) = tr^2(g) \text{ and } Tr(T) = |tr(g)|$$

are well-defined functions of T. An important geometrical meaning of the *trace* function Tr(T) will be discussed in §2.1. Notice that $PSL(2,\mathbb{R})$ contains all fractional linear transformations of the form $z \to \frac{az+b}{cz+d}$ with $a,b,c,d \in \mathbb{R}$ and $\Delta = ad - bc > 0$ since by dividing the numerator and denominator by $\sqrt{\Delta}$ we obtain a new matrix for it of determinant 1. In particular, $PSL(2,\mathbb{R})$ contains all transformations of the form $z \to az + b$ ($a,b \in \mathbb{R}$, $a > 0$), and the transformation $z \to -\frac{1}{z}$.

THEOREM 1.1.1. $PSL(2,\mathbb{R})$ *acts on* $\mathcal{H}$ *by homeomorphisms.*

PROOF: First we show that any transformation (1.1.2) maps $\mathcal{H}$ onto itself. Let $T \in PSL(2,\mathbb{R})$, and $w = T(z) = \frac{az+b}{cz+d}$. Then

$$w = \frac{(az+b)(c\bar{z}+d)}{|cz+d|^2} = \frac{ac|z|^2 + adz + bc\bar{z} + bd}{|cz+d|^2},$$

so that

$$\mathrm{Im}(w) = \frac{w - \bar{w}}{2i} = \frac{z - \bar{z}}{2i|cz+d|^2} = \frac{\mathrm{Im}(z)}{|cz+d|^2}. \tag{1.1.3}$$

Therefore $\mathrm{Im}(z) > 0$ implies $\mathrm{Im}(w) > 0$. The theorem now follows from the continuity of T(z) and its inverse. □

DEFINITION. A transformation of $\mathcal{H}$ onto itself is called an *isometry* if it preserves the hyperbolic distance on $\mathcal{H}$.

It is clear that the set of all isometries of $\mathcal{H}$ forms a group; we shall denote it by Isom($\mathcal{H}$).

THEOREM 1.1.2. $PSL(2,\mathbb{R}) \subset \mathrm{Isom}(\mathcal{H})$.

PROOF: By Theorem 1.1.1 all transformations in $PSL(2,\mathbb{R})$ map $\mathcal{H}$ onto itself. We show that, if $\gamma: I \to \mathcal{H}$ is a piecewise differentiable path in $\mathcal{H}$ then for any $T \in PSL(2,\mathbb{R})$ we have $h(T(\gamma)) = h(\gamma)$. Suppose $\gamma: I \to \mathcal{H}$ is given by $z(t) = (x(t), y(t))$, and $w(t) = T(z(t)) = u(t) + iv(t)$. We have

$$\frac{dw}{dz} = \frac{a(cz+d) - c(az+b)}{(cz+d)^2} = \frac{1}{(cz+d)^2}. \tag{1.1.4}$$

By (1.1.3) $v = \frac{y}{|cz+d|^2}$, and hence $\left|\frac{dw}{dz}\right| = \frac{v}{y}$. Thus

$$h(T(\gamma)) = \int_0^1 \frac{\left|\frac{dw}{dt}\right| dt}{v(t)} = \int_0^1 \frac{\left|\frac{dw}{dz}\frac{dz}{dt}\right| dt}{v(t)} = \int_0^1 \frac{\left|\frac{dz}{dt}\right| dt}{y(t)} = h(\gamma).$$

The invariance of the hyperbolic distance follows at once from the above invariance. □

1.2. Geodesics

THEOREM 1.2.1. *The geodesics in $\mathcal{H}$ are semicircles and straight lines orthogonal to the real axis $\mathbb{R}$.*

PROOF: Let z_1 and z_2 be two point in $\mathcal{H}$. Suppose first that $z_1 = ia$, $z_2 = ib$ $(b > a)$. If $\gamma: I \to \mathcal{H}$ is any piecewise differentiable path joining ia and ib, with $\gamma(t) = (x(t), y(t))$, then

$$h(\gamma) = \int_0^1 \frac{\sqrt{\left(\frac{dx}{dt}\right)^2 + \left(\frac{dy}{dt}\right)^2}\, dt}{y(t)} \geq \int_0^1 \frac{\left|\frac{dy}{dt}\right| dt}{y(t)} \geq \int_0^1 \frac{\frac{dy}{dt} dt}{y(t)} = \int_a^b \frac{dy}{y} = \ln \frac{b}{a}, \tag{1.2.1}$$

but $\ln \frac{b}{a}$ is the hyperbolic length of the segment of the y-axis joining ia and ib, hence the geodesic joining ia and ib is the segment of the imaginary axis joining them. For arbitrary z_1 and z_2 in $\mathcal{H}$, let L be the unique Euclidean circle or straight line orthogonal to the real axis $\mathbb{R}$ passing through those points. By Exercise 1.1 there exists a transformation in PSL(2,$\mathbb{R}$) which maps L into the imaginary axis. Using the above argument and Theorem 1.1.2 we conclude that the geodesic joining z_1 and z_2 is the segment of L joining them. □

COROLLARY 1.2.2. *Any two points* z, w$\in\mathcal{H}$ *can be joined by a unique geodesic, and the hyperbolic distance between* z, w$\in\mathcal{H}$ *is equal to the hyperbolic length of the unique geodesic segment connecting these two points which we denote* [z, w].

COROLLARY 1.2.3. *If* z *and* w *are two distinct points in* $\mathcal{H}$, *then*

$$\rho(z, w)=\rho(z, \xi)+\rho(\xi, w)$$

if and only if $\xi\in[z,w]$.

THEOREM 1.2.4. *Any transformation in* PSL(2,$\mathbb{R}$) *maps geodesics onto geodesics in* $\mathcal{H}$.

PROOF: Let T$\in$PSL(2,$\mathbb{R}$), z and w be two distinct points in $\mathcal{H}$, and $\xi\in[z,w]$. By Theorem 1.1.2 and Corollary 1.2.3, $T\xi\in[Tz, Tw]$, i.e. T maps the segment [z,w] onto the segment [Tz, Tw], and hence geodesics to geodesics. We shall also give an independent proof based only on the geometrical description of geodesics in $\mathcal{H}$. We know from complex analysis that fractional linear transformations map Euclidean lines and circles to lines and circles. T transforms the real axis onto itself, and hence the lines and circles orthogonal to the real axis onto lines and circles orthogonal to the real axis. □

It is useful for many purposes to extend the complex plane $\mathbb{C}$ by introduction of a symbol ∞ to represent infinity; $\hat{\mathbb{C}}=\mathbb{C}\cup\{\infty\}$ is also called the *Riemann sphere.* The *cross-ratio* of distinct points z_1, z_2, z_3, $z_4\in\hat{\mathbb{C}}$ is given by the formula

$$(z_1, z_2 ; z_3, z_4)=\frac{(z_1-z_2)(z_3-z_4)}{(z_2-z_3)(z_4-z_1)}.$$

THEOREM 1.2.5. *Let* $z,w\in\mathcal{H}$ $(z\neq w)$ *and let the geodesic joining* z *and* w *have endpoints* z^*,w^* *in* $\mathbb{R}\cup\{\infty\}$, *chosen in such a way that* z *lies between* z^* *and* w. *Then*

$$\rho(z, w)=\ln (w, z^*; z, w^*).$$

PROOF: According to Exercise 1.1 there exists an element $T\in PSL(2,\mathbb{R})$ which maps the geodesic joining z and w to the imaginary axis. By applying transformations $z\rightarrow kz$ $(k>0)$ and $z\rightarrow -\frac{1}{z}$ as necessary, we may assume that $T(z^*)=0$, $T(w^*)=\infty$ and $T(z)=i$. Then $T(w)=ri$ $(r>1)$, and by (1.2.1) $\rho(z,w)=\ln r$. But $r=(ri,0; i,\infty)$, and the Theorem follows from the invariance of the cross-ratio under fractional linear transformations (see e.g. [A]). □

We shall now derive explicit formulae for the hyperbolic distance.

THEOREM 1.2.6. *For* $z,w\in\mathcal{H}$,

(i) $$\rho(z, w)=\ln \frac{|z-\bar{w}|+|z-w|}{|z-\bar{w}|-|z-w|};$$

(ii) $$\cosh \rho(z, w)=1+\frac{|z-w|^2}{2\,\mathrm{Im}(z)\,\mathrm{Im}(w)};$$

(iii) $$\sinh \left[\tfrac{1}{2}\rho(z, w)\right]=\frac{|z-w|}{2\,(\mathrm{Im}(z)\,\mathrm{Im}(w))^{1/2}};$$

(iv) $$\cosh\left[\tfrac{1}{2}\rho(z, w)\right]=\frac{|z-\bar{w}|}{2\,(\mathrm{Im}(z)\,\mathrm{Im}(w))^{1/2}};$$

(v) $$\tanh\left[\tfrac{1}{2}\rho(z, w)\right]=\left|\frac{z-w}{z-\bar{w}}\right|.$$

PROOF: It is a routine exercise to check that the five equations are equivalent to each other; we shall prove that (iii) holds. By Theorem 1.1.2, the left-hand side of (iii) is invariant under every $T\in PSL(2, \mathbb{R})$. Exercise 1.4 shows that the right-hand side is also invariant under T. Let L be the unique geodesic passing through z and w, and T_0 be a transformation mapping L to the imaginary axis as in Exercise 1.1. It is now only necessary to check (iii) when z and w lie on the imaginary axis: $z=ia$, $w=ib$ $(a<b)$. We have seen in the proof of Theorem 1.1.2 that $\rho(ia, ib)=\ln\frac{b}{a}$, and it is easy to see that (iii) holds in this case. □

We shall now describe a model of the hyperbolic geometry in the *unit disc:*

$$\mathcal{U}=\{z\in\mathbb{C},\ |z|<1\}.$$

The map

$$f(z)=\frac{zi+1}{z+i} \qquad (1.2.2)$$

is a 1-1 map of $\mathcal{H}$ onto $\mathcal{U}$, thus ρ^* given by

$$\rho^*(z,w)=\rho(f^{-1}z,\ f^{-1}w) \quad (z,w\in\mathcal{U})$$

is a metric on $\mathcal{U}$. Using Exercise 1.5 we see that ρ^* can be identified with the metric derived from the differential

$$ds=\frac{2|dz|}{1-|z|^2}. \qquad (1.2.3)$$

We prefer to use ρ for ρ^* and with this convention f *is an isometry of* $(\mathcal{H},\rho)$ *onto* $(\mathcal{U},\rho)$. We shall refer to these two models of hyperbolic geometry as the Poincaré models, and we shall change from one model to the other as each has its own particular advantage.

The circle $\Sigma=\{z\in C \mid |z|=1\}$ is called the *principal circle;* it is the *Euclidean boundary* of $\mathcal{U}$. Similarly, the *Euclidean boundary* of the upper half-plane $\mathcal{H}$, considered as a subset of the Riemann sphere $\hat{C}$, is $\mathbb{R}\cup\{\infty\}$. In the model $\mathcal{U}$ geodesics are segments of Euclidean circles orthogonal to the principal circle Σ and its diameters (see Exercise 1.7). For the formulae for the hyperbolic distance in $\mathcal{U}$ analogous to those of Theorem 1.2.6 see Exercise 1.8.

Formula (1.2.1) shows that the set of points in the Euclidean boundary of the hyperbolic plane is characterized by the property that the hyperbolic distance from these points to any point in the hyperbolic plane is infinite; we shall sometimes refer to this set as *points at infinity.*

Let $\widetilde{\mathcal{H}}=\mathcal{H}\bigcup\mathbb{R}\bigcup\{\infty\}$ be the *Euclidean closure of* $\mathcal{H}$, and $\widetilde{\mathcal{U}}=\mathcal{U}\bigcup\Sigma$ be the *Euclidean closure* of $\mathcal{U}$. We see that the Euclidean closure of $\mathcal{H}$ (resp. $\mathcal{U}$) is the closure of $\mathcal{H}$ (resp. $\mathcal{U}$) in $\hat{C}$.

1.3. Isometries

We have seen (Theorem 1.1.2) that the transformations in $PSL(2,\mathbb{R})$ are *isometries* of $\mathcal{H}$. Let $PS^*L(2,\mathbb{R})=S^*L(2,\mathbb{R})/\{\pm 1_2\}$ where $S^*L(2,\mathbb{R})$ is a group of real matrices $g=\begin{bmatrix} a & b \\ c & d \end{bmatrix}$ with $\det(g)=\pm 1$. $PS^*L(2,\mathbb{R})$ contains the group $PSL(2,\mathbb{R})$ as a subgroup of index 2.

The following Theorem identifies <u>all</u> isometries of the hyperbolic plane $\mathcal{H}$.

THEOREM 1.3.1. *The group* $\text{Isom}(\mathcal{H})$ *is generated by the fractional* *linear transformations* (1.1.2) *in* PSL(2,$\mathbb{R}$) *together with* $z \to -\bar{z}$, *and is isomorphic to* PS*L(2,$\mathbb{R}$). *The group* PSL(2,$\mathbb{R}$) *is a subgroup of* Isom($\mathcal{H}$) *of index* 2.

PROOF: Let ϕ be any isometry of $\mathcal{H}$. Repeating the argument of the first proof of Theorem 1.2.4 we conclude that ϕ maps geodesics to geodesics. Let I denote the positive imaginary axis. $\phi(I)$ is a geodesic, and according to Exercise 1.1 there exists an isometry $g \in PSL(2,\mathbb{R})$ which maps $\phi(I)$ to I. By applying transformations $z \to kz$ $(k>0)$ and $z \to -\frac{1}{z}$ as we did in the proof of Theorem 1.2.5, we may assume that $g\phi$ fixes i and maps the rays (i, ∞) and $(0, i)$ onto themselves, and hence $g\phi$ fixes each point of I.

Now let $z=x+iy \in \mathcal{H}$, and $g\phi(z)=u+iv$. For all positive t,

$$\rho(z,it)=\rho(g\phi(z),g\phi(it))=\rho(u+iv,it)$$

and by Theorem 1.2.6(iii),

$$[x^2+(y-t)^2]v=[u^2+(v-t)^2]y.$$

As this holds for all positive t, dividing the both sides of the above equation by t^2 and taking the limit as $t \to \infty$, we have $v=y$, and $x^2=u^2$. Thus

$$g\phi(z)=z \text{ or } -\bar{z}. \tag{1.3.1}$$

Since isometries are continuous (see Exercise 1.9) only one of the equations (1.3.1) holds for all z in $\mathcal{H}$. If $g\phi(z)=z$, then $\phi(z)$ is a fractional linear transformation of the form (1.1.2). If $g\phi(z)=-\bar{z}$, we have

$$\phi(z)=\frac{a\bar{z}+b}{c\bar{z}+d} \text{ with } ad-bc=-1. \tag{1.3.2}$$

Thus we have identified all isometries of $\mathcal{H}$. It is easy to check that all transformations of the form (1.1.2) and (1.3.2) form a group which is isomorphic to the group $PS^*L(2,\mathbb{R})$. The sign of the determinant of the corresponding matrix $\begin{bmatrix} a & b \\ c & d \end{bmatrix}$ determines the *orientation* of an isometry; thus transformations in $PSL(2,\mathbb{R})$ are *orientation-preserving* while transformations of the form (1.3.2), in particular, $z \to -\bar{z}$, are *orientation-reversing isometries*. □

Let us consider now the tangent space to $\mathcal{H}$ at a point z, $T_z\mathcal{H} \approx \mathbb{C}$. The Riemannian metric (1.1.1) on $\mathcal{H}$ is induced by the following inner product on $T_z\mathcal{H}$: for $\zeta_1 = \xi_1 + i\eta_1$ and $\zeta_2 = \xi_2 + i\eta_2$ in $T_z\mathcal{H}$

$$<\zeta_1, \zeta_2> = \frac{1}{(\mathrm{Im}z)^2}(\xi_1\xi_2 + \eta_1\eta_2). \tag{1.3.3}$$

We shall denote the norm in $T_z\mathcal{H}$ corresponding to this inner product by $\|\cdot\|$. Since isometries of $\mathcal{H}$ (being transformations of the form (1.1.2) or (1.3.2)) are differentiable mappings, they act on the tangent bundle $T\mathcal{H}$ by differentials preserving the norm in the tangent bundle (Exercise 1.2). By the polarization identity, for any $\xi, \eta \in T_z\mathcal{H}$

$$<\xi,\eta> = \tfrac{1}{2}(\|\xi\|^2 + \|\eta\|^2 - \|\xi - \eta\|^2);$$

thus the inner product and hence the absolute value of an *angle* between tangent vectors are also preserved. We define an *angle* between two geodesics in $\mathcal{H}$ at their intersection point z as the angle between their tangent vectors in $T_z\mathcal{H}$. By Exercise 1.3 this notion of an angle coincides with the Euclidean angle measure.

DEFINITION. A transformation of $\mathcal{H}$ is called *conformal* if it preserves angles, and *anti-conformal* if it preserves the absolute values of angles but changes the signs.

THEOREM 1.3.2. *Any transformation of* PSL(2,$\mathbb{R}$) *is conformal; any transformation of the form* (1.3.2) *is anti-conformal.*

PROOF: This follows immediately from the above discussion and Theorem 1.3.1. Alternatively, any $T(z)=\frac{az+b}{cz+d}\in PSL(2,\mathbb{R})$ is conformal on the complex plane since for any $z\in\mathbb{C}$, $T'(z)=\frac{1}{(cz+d)^2}\neq 0$, and the transformation $z\to -\bar{z}$ is obviously anti-conformal. □

It follows from Exercise 1.11 that the family of all Euclidean discs coincides with the family of all hyperbolic discs, and we have the following result.

THEOREM 1.3.3. *The topology on* $\mathcal{H}$ *induced by the hyperbolic metric is the same as the topology induced by the Euclidean metric.*

1.4. Hyperbolic area and the Gauss - Bonnet formula

For a subset $A\subseteq\mathcal{H}$ we define $\mu(A)$, the *hyperbolic area* of A, by

$$\mu(A)=\int_A \frac{dxdy}{y^2} \tag{1.4.1}$$

if this integral exists.

THEOREM 1.4.1. *The hyperbolic area is invariant under all transformations in* PSL(2,$\mathbb{R}$): *if* $A\subseteq\mathcal{H}$, $\mu(A)$ *exists, and* $T\in PSL(2,\mathbb{R})$ *then* $\mu(T(A))=\mu(A)$.

PROOF: Let $z=x+iy$,

$$T(z)=\frac{az+b}{cz+d} \quad (a,b,c,d\in\mathbb{R},\ ad-bc=1),$$

and $w=T(z)=u+iv$. Then using the Cauchy-Riemann equations we calculate the Jacobian

$$\frac{\partial(u,v)}{\partial(x,y)}=\frac{\partial u}{\partial x}\frac{\partial v}{\partial y}-\frac{\partial u}{\partial y}\frac{\partial v}{\partial x}=\left(\frac{\partial u}{\partial x}\right)^2+\left(\frac{\partial v}{\partial x}\right)^2=\left|\frac{dT}{dz}\right|^2=\frac{1}{|cz+d|^4}.$$

Thus

$$\mu(T(A))=\int_{T(A)}\frac{du\,dv}{v^2}=\int_A\frac{\partial(u,v)}{\partial(x,y)}\frac{dx\,dy}{v^2}=\int_A\frac{1}{|cz+d|^4}\frac{|cz+d|^4}{y^2}dx\,dy=\mu(A)$$

using (1.1.3). □

A *hyperbolic* n-*sided polygon* is a closed set of $\bar{\mathcal{H}}$ bounded by n hyperbolic geodesic segments. If two line segments intersect, then the point of their intersection is called a *vertex* of the polygon. We allow vertices on $\mathbb{R}\cup\{\infty\}$ although no segment of the real axis can belong to a hyperbolic polygon.

In Figure 2 we illustrate four types of hyperbolic triangles depending on whether 0, 1, 2, or 3 vertices of the triangle belong to $\mathbb{R}\cup\{\infty\}$.

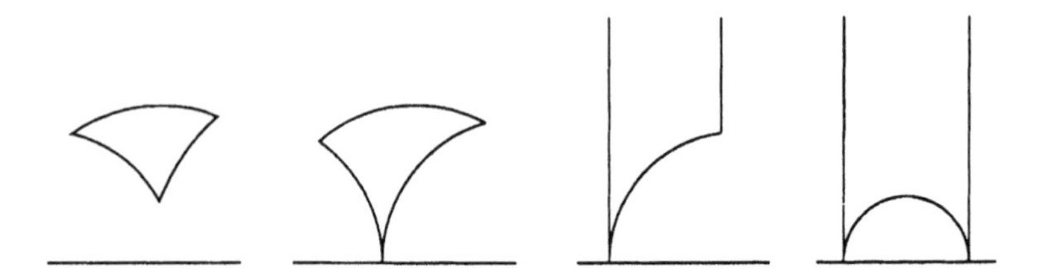

Fig. 2

The Gauss–Bonnet formula shows that the hyperbolic area of a hyperbolic triangle depends only on its angles.

THEOREM 1.4.2. (Gauss–Bonnet) *Let Δ be a hyperbolic triangle with angles α,β,γ. Then*

$$\mu(\Delta)=\pi-\alpha-\beta-\gamma.$$

PROOF: *Case 1.* We suppose first that one of the vertices of Δ belongs to $\mathbb{R}\cup\{\infty\}$, and hence the angle at this vertex is equal to zero. If it belongs to the real axis, by applying a transformation in PSL(2,$\mathbb{R}$) we can map this vertex to ∞ without altering the hyperbolic area or the angles. It is sufficient therefore to consider the case where two sides of Δ are vertical geodesics. The base of Δ is then a segment of a Euclidean semicircle orthogonal to the real axis. By applying transformations of the form $z\to z+k$ ($k\in\mathbb{R}$), $z\to\lambda z$ ($\lambda>0$) we can assume that the semicircle has center 0 and radius 1 (see Fig. 3). These transformations will not change the area of Δ, by Theorem 1.4.1. The zero angle will be preserved since these transformations map vertical geodesics to vertical geodesics; the other angles will be preserved by conformality (Theorem 1.3.2).

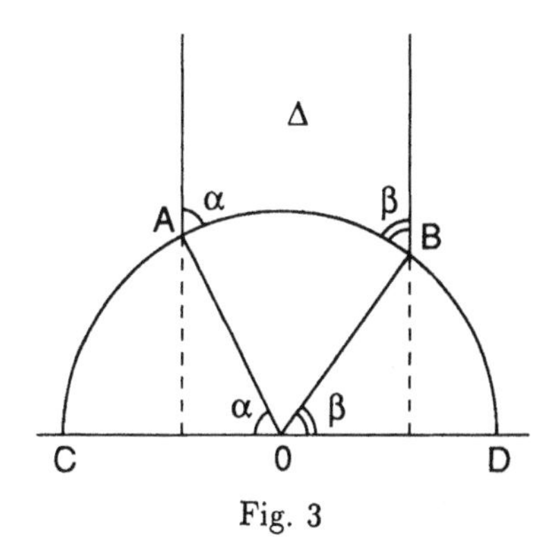

Fig. 3

The angles AOC and BOD are equal to α and β, respectively, as angles with mutually perpendicular sides. Assume that the vertical geodesics through A and B are the lines x=a and x=b, respectively. We now calculate

$$\mu(\Delta)=\int_{\Delta}\frac{dxdy}{y^2}=\int_a^b dx\int_{\sqrt{1-x^2}}^{\infty}\frac{dy}{y^2}=\int_a^b\frac{dx}{\sqrt{1-x^2}}.$$

Make the substitution $x=\cos\theta$ $(0\leq\theta\leq\pi)$; then

$$\mu(\Delta)=\int_{\pi-\alpha}^{\beta}\frac{-\sin\theta d\theta}{\sin\theta}=\pi-\alpha-\beta.$$

Case 2. Δ has no vertices in $\mathbb{R}\cup\{\infty\}$. Suppose that Δ has vertices A,B,C and that the geodesic connecting A and B intersects the real axis at D. (We can apply a transformation in PSL(2,$\mathbb{R}$) to make sure that no sides of Δ are vertical geodesics.) Then we have the situation of Figure 4. Here $\Delta=\Delta_1-\Delta_2$ where Δ_1 has vertices A,C,D and Δ_2 has vertices B,C,D.

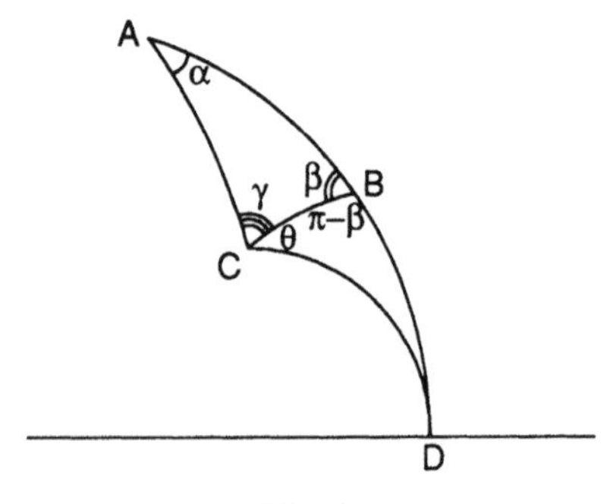

Fig. 4

Then

$$\mu(\Delta)=\mu(\Delta_1)-\mu(\Delta_2)=\pi-\alpha-(\gamma+\theta)-[\pi-\theta-(\pi-\beta)]=\pi-\alpha-\beta-\gamma.$$

□

1.5. Hyperbolic trigonometry

The *angle of parallelism* is a classical term for the angle α determined by the trigonometric relation which holds for a triangle with angles α, 0, $\frac{\pi}{2}$ ($\alpha\neq 0$). Let us denote the length of the only finite side of the triangle by a (see Fig. 5). The angle of parallelism α is a function of a: $\alpha=\Pi(a)$.

THEOREM 1.5.1. *Let Δ be a triangle with angles* 0 *and* $\frac{\pi}{2}$ *and the finite side* a. *Then for the third angle* $\Pi(a)$ *we have*

$$\text{(i)}\quad \tan\Pi(a)=\frac{1}{\sinh a},$$
$$\text{(ii)}\quad \sin\Pi(a)=\frac{1}{\cosh a},$$
$$\text{(iii)}\quad \sec\Pi(a)=\frac{1}{\tanh a}.$$

PROOF: We shall prove (ii). By Theorem 1.2.6(ii), we have

$$\cosh a=1+\frac{4\sin^2(\frac{\pi}{4}-\frac{\Pi(a)}{2})}{2\sin\Pi(a)}=\frac{\sin\Pi(a)+2\sin^2(\frac{\pi}{4}-\frac{\Pi(a)}{2})}{\sin\Pi(a)}=\frac{1}{\sin\Pi(a)}.$$

The remaining formulae are equivalent to (ii). □

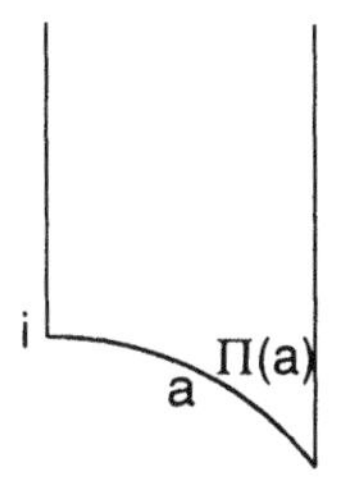

Fig. 5

We now consider the general hyperbolic triangle with sides of hyperbolic length a, b, c and opposite angles α, β, γ. We assume that α, β, and γ are positive (so a, b, and c are finite) and prove the following results.

THEOREM 1.5.2.

(i) The Sine Rule: $\dfrac{\sinh a}{\sin\alpha}=\dfrac{\sinh b}{\sin\beta}=\dfrac{\sinh c}{\sin\gamma}$.

(ii) The Cosine Rule I: $\cosh c=\cosh a \cosh b-\sinh a \sinh b \cos\gamma$.

(iii) The Cosine Rule II: $\cosh c=\dfrac{\cos\alpha \cos\beta+\cos\gamma}{\sin\alpha \sin \beta}$.

REMARK: Note the existence of Cosine Rule II. This has no analogue in Euclidean geometry: in hyperbolic geometry it implies that *if two triangles have the same angles, then there is an isometry mapping one triangle onto the other.*

PROOF of (ii): Let us denote the vertices opposite the sides a, b, c by v_a, v_b, v_c respectively.

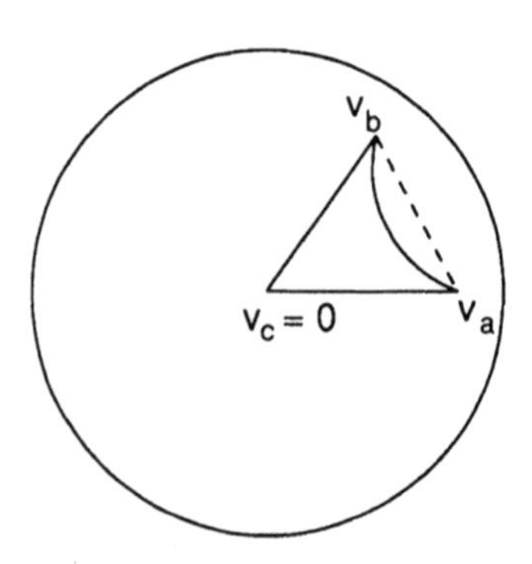

Fig. 6

We shall use the model $\mathcal{U}$ and may assume that $v_c=0$ and Im $v_a=0$, Re $v_a>0$ (see Fig. 6). By Exercise 1.8 (iv) we have

$$v_a=\tanh\tfrac{1}{2}\rho(0,v_a)=\tanh(\tfrac{1}{2}b), \tag{1.5.1}$$

and similarly,

$$v_b=e^{i\gamma}\tanh(\tfrac{1}{2}a). \tag{1.5.2}$$

We have $c=\rho(v_a,v_b)$, and from Exercise 1.8 (iii)

$$\cosh c=2\sinh^2[\tfrac{1}{2}\rho(v_a,\ v_b)]+1=\frac{2|v_a-v_b|^2}{(1-|v_a|^2)(1-|v_b|^2)}+1. \tag{1.5.3}$$

The right-hand side of expression (1.5.3) is equal to cosh a cosh b−sinh a sinh b cosγ by Exercise 1.12, and hence (ii) follows. □

PROOF of (i): Using (ii) we obtain

$$\left(\frac{\sinh c}{\sin\gamma}\right)^2=\frac{\sinh^2 c}{1-\left(\dfrac{\cosh a\cosh b-\cosh c}{\sinh a\sinh b}\right)^2}. \tag{1.5.4}$$

The Sine Rule will be valid if we prove that the expression on the right-hand side of (1.5.4) is symmetric in a, b, and c. This follows from the symmetry of

$$(\sinh a\sinh b)^2-(\cosh a\cosh b-\cosh c)^2$$

which is obtained by a direct calculation. □

PROOF of (iii): Let us write A for cosh a, B for cosh b, and C for cosh c. The Cosine Rule I yields

$$\cos\gamma=\frac{(AB-C)}{(A^2-1)^{\frac{1}{2}}(B^2-1)^{\frac{1}{2}}}$$

and so

$$\sin^2\gamma = \frac{D}{(A^2-1)(B^2-1)},$$

where $D=1+2ABC-(A^2+B^2+C^2)$ is symmetric in A, B, and C. The expression for $\sin^2\gamma$ shows that $D \geq 0$. Using analogous expressions for $\cos\alpha$, $\sin\alpha$, $\cos\beta$, and $\sin\beta$ we observe that if we multiply both the numerator and denominator of

$$\frac{\cos\alpha \cos\beta + \cos\gamma}{\sin\alpha \sin\beta}$$

by the positive value of

$$(A^2-1)^{\frac{1}{2}}(B^2-1)^{\frac{1}{2}}(C^2-1),$$

we obtain

$$\frac{\cos\alpha \cos\beta + \cos\gamma}{\sin\alpha \sin\beta} = \frac{[(BC-A)(CA-B)+(AB-C)(C^2-1)]}{D} = C.$$

□

THEOREM 1.5.3. (Pythagorian Theorem) *If* $\gamma = \frac{\pi}{2}$ *we have* $\cosh c = \cosh a \cosh b$.

PROOF: Immediate from the Cosine Rule I. □

1.6. Comparison between hyperbolic, spherical and Euclidean trigonometry

Let S_r be a sphere of radius r in $\mathbb{R}^3$, and $\mathcal{H}_r$ be the upper half-plane equipped with the metric

$$ds = \frac{r\sqrt{dx^2+dy^2}}{y}. \tag{1.6.1}$$

$\mathcal{H}_r$ is a model of a *sphere of imaginary radius* ir.

S_r is a manifold of positive curvature $\frac{1}{r^2}$, and $\mathcal{H}_r$ is a manifold of negative curvature $-\frac{1}{r^2}$. The geodesics in S_r are the great circles.

We shall denote the angles of a triangle by α, β, γ, and the lengths of the opposite sides by a, b, and c.

We have the following trigonometric formulae for a *spherical triangle in* S_r.

(i) The Sine Rule: $\frac{\sin\frac{a}{r}}{\sin\alpha}=\frac{\sin\frac{b}{r}}{\sin\beta}=\frac{\sin\frac{c}{r}}{\sin\gamma}$.

(ii) The Cosine Rule I: $\cos\frac{c}{r}=\cos\frac{a}{r}\cos\frac{b}{r}+\sin\frac{a}{r}\sin\frac{b}{r}\cos\gamma$. (1.6.2)

(iii) The Cosine Rule II: $\cos\frac{c}{r}=\frac{\cos\alpha\cos\beta+\cos\gamma}{\sin\alpha\sin\beta}$.

Plugging into these formulae ir instead of r and using the identities

$$\sin ix=i\sinh x,\ \cos ix=\cosh x,$$

we obtain trigonometric formulae for a *hyperbolic triangle in* $\mathcal{H}_r$ (compare with Theorem 1.5.2 for $\mathcal{H}=\mathcal{H}_1$).

(i) The Sine Rule: $\frac{\sinh\frac{a}{r}}{\sin\alpha}=\frac{\sinh\frac{b}{r}}{\sin\beta}=\frac{\sinh\frac{c}{r}}{\sin\gamma}$.

(ii) The Cosine Rule I: $\cosh\frac{c}{r}=\cosh\frac{a}{r}\cosh\frac{b}{r}-\sinh\frac{a}{r}\sinh\frac{b}{r}\cos\gamma$. (1.6.3)

(iii) The Cosine Rule II: $\cosh\frac{c}{r}=\frac{\cos\alpha\cos\beta+\cos\gamma}{\sin\alpha\sin\beta}$.

In order to obtain the formulae of Euclidean trigonometry we replace for large r and x=a, b, c, $\sin\frac{x}{r}$ and $\sinh\frac{x}{r}$ with $\frac{x}{r}$, $\cos\frac{x}{r}$ with $1-\frac{x^2}{2r^2}$, and $\cosh\frac{x}{r}$ with $1+\frac{x^2}{2r^2}$ in formulae (1.6.2) and (1.6.3), and let

$r \to \infty$. Thus we obtain

(i) The Sine Rule: $\frac{a}{\sin\alpha} = \frac{b}{\sin\beta} = \frac{c}{\sin\gamma}$.

(ii) The Cosine Rule: $c^2 = a^2 + b^2 - 2ab\cos\gamma$. (1.6.4)

(iii) $\cos\gamma = -\cos\cos\alpha\cos\beta + \sin\alpha\sin\beta = -\cos(\alpha+\beta)$,
which is equivalent to $\alpha+\beta+\gamma=\pi$.

EXERCISES FOR CHAPTER 1

1.1. Let L be a Euclidean circle or a straight line orthogonal to the real axis which meets the real axis at some finite point α. Prove that the transformation $T(z)=-(z-\alpha)^{-1}+\beta$ belongs to PSL(2,$\mathbb{R}$), and for a suitable β maps L to the imaginary axis.

1.2. Prove that a differentiable mapping of $\mathcal{H}$ onto itself is an isometry if and only if its differential preserves the norm on the tangent bundle of $\mathcal{H}$.

1.3. Show that the notion of an angle introduced in §1.2 is the same as the Euclidean angle measure.

1.4. Prove that for $z,w\in\mathcal{H}$ and $T\in PSL(2,\mathbb{R})$, $|Tz-Tw|=|z-w||T'(z)T'(w)|^{1/2}$.

1.5. Prove that for $z\in\mathcal{H}$ and $f(z)=\frac{zi+1}{z+i}$, $\frac{2|f'(z)|}{1-|f(z)|^2}=\frac{1}{\mathrm{Im}(z)}$.

1.6. Prove that in model $\mathcal{U}$, if $0<r<1$ then $\rho(0,ir)=\int_0^r\frac{2dt}{1-t^2}=\ln\frac{1+r}{1-r}$.

1.7. Prove that geodesics in $\mathcal{U}$ are segments of Euclidean circles orthogonal to the principal circle Σ and its diameters.

1.8. Rewrite Theorem 1.2.6 by means of the map f (1.2.2). Prove the following formulae for z, w $\in\mathcal{U}$:

(i) $$\rho(z,w)=\ln\frac{|1-z\bar{w}|+|z-w|}{|1-z\bar{w}|-|z-w|},$$

(ii) $$\cosh^2[\tfrac{1}{2}\rho(z,w)]=\frac{|1-z\bar{w}|^2}{(1-|z|^2)(1-|w|^2)},$$

(iii) $$\sinh^2[\tfrac{1}{2}\rho(z,w)]=\frac{|z-w|^2}{(1-|z|^2)(1-|w|^2)},$$

(iv) $$\tanh[\tfrac{1}{2}\rho(z,w)]=\left|\frac{z-w}{1-z\bar{w}}\right|.$$

1.9. Prove that isometries are continuous.

1.10. Show that the group of orientation-preserving isometries of $\mathcal{U}$

is the group of fractional linear transformations of the form

$$z \rightarrow \frac{az+\bar{c}}{cz+\bar{a}} \quad (a,c \in \mathbb{C},\ a\bar{a}-c\bar{c}=1).$$

1.11. Show that every hyperbolic circle in $\mathcal{H}$ is a Euclidean circle (with a different center, of course), and vice versa.

1.12 Show that the right-hand side of expression (1.5.3) is equal to $\cosh a \cosh b - \sinh a \sinh b \cos\gamma$.

1.13. (J.Bolyai) Prove that the Sine Rule can be written in the form

$$\frac{\sin\alpha}{\odot a} = \frac{\sin\beta}{\odot b} = \frac{\sin\gamma}{\odot c},$$

with $\odot r$ being the circumference of the circle of radius r, which is valid in both hyperbolic and Euclidean geometries.

2. FUCHSIAN GROUPS

2.1. The group PSL(2, $\mathbb{R}$)

There are three types of elements in

$$PSL(2, \mathbb{R})=\left\{z\to T(z)=\frac{az+b}{cz+d} \mid ad-bc=1\right\}$$

distiguished by the value of its *trace:* $Tr(T)=|a+d|$. If $Tr(T)<2$, T is called *elliptic;* if $Tr(T)=2$, T is called *parabolic;* and if $Tr(T)>2$, T is called *hyperbolic.*

The terminology can be explained by looking at the linear action of the corresponding matrices on $\mathbb{R}^2$. A matrix in SL(2, $\mathbb{R}$) is hyperbolic if and only if it is diagonalizable over $\mathbb{R}$, or conjugate in SL(2, $\mathbb{R}$) to a unique matrix $\begin{bmatrix} \lambda & 0 \\ 0 & 1/\lambda \end{bmatrix}$, $\lambda\neq 1$, and it is elliptic if and only if it is conjugate in SL(2, $\mathbb{R}$) to a unique matrix $\begin{bmatrix} \cos\theta & \sin\theta \\ -\sin\theta & \cos\theta \end{bmatrix}$. It follows that the invariant curves for hyperbolic (resp. elliptic) linear transformations of $\mathbb{R}^2$ are *hyperbolas* (resp. *ellipses*), hence the terminology for hyperbolic (resp. elliptic) transformations. The parabolic transformations are so called by analogy, as intermediate between hyperbolic and elliptic.

The fixed points are found by solving

$$z=\frac{az+b}{cz+d} \quad (a, b, c, d\in\mathbb{R}, ad-bc=1),$$

and we see that a hyperbolic transformation has two fixed points in $\mathbb{R}\cup\{\infty\}$, one *repulsive* and one *attractive*, a parabolic transformation has one fixed point in $\mathbb{R}\cup\{\infty\}$, and an elliptic transformation has a pair of complex conjugate fixed points, and therefore, one fixed point in $\mathcal{H}$. A transformation (1.1.2) fixes ∞ if and only if $c=0$ and hence it is of the form $z \to az+b$ ($a,b \in \mathbb{R}$, $a>0$). If $a=1$, it is parabolic; if $a \neq 1$, it is hyperbolic and its second fixed point is $\frac{b}{1-a}$.

DEFINITION: A geodesic in $\mathcal{H}$ joining the two fixed points of a hyperbolic transformation T is called the *axis* of T, and is denoted by C(T).

By Theorem 1.2.5, T maps C(T) onto itself.

Let $S\mathcal{H}$ be the unit tangent bundle of the upper half-plane $\mathcal{H}$. It is homeomorphic to $\mathcal{H} \times S^1$. Let us parametrize it by local coordinates (z,ζ), where $z \in \mathcal{H}$, $\zeta \in \mathbb{C}$ with $|\zeta| = \mathrm{Im}(z)$. (Notice that with this parametrization, $\|\zeta\|=1$ (see (1.3.3)), so that ζ is a unit tangent vector.) It is easy to check that the group PSL(2,$\mathbb{R}$) acts on $S\mathcal{H}$ by the differentials: for $T : z \to \frac{az+b}{cz+d}$, $T(z,\zeta)=(T(z), DT(\zeta))$, where

$$DT(\zeta)=\frac{1}{(cz+d)^2}\zeta \ . \tag{2.1.1}$$

As any group, PSL(2,$\mathbb{R}$) acts on itself by left multiplication. The next result connects these two actions, and is an elaboration of Exercise 2.3 which implies that PSL(2,$\mathbb{R}$) and $S\mathcal{H}$ are homeomorphic as topological spaces.

THEOREM 2.1.1. *There is a homeomorphism between* PSL(2,$\mathbb{R}$) *and the unit tangent bundle* $S\mathcal{H}$ *of the upper half-plane* $\mathcal{H}$ *such that the action of* PSL(2,$\mathbb{R}$) *on itself by left multiplication corresponds to the action of* PSL(2,$\mathbb{R}$) on $S\mathcal{H}$ *induced by its action on* $\mathcal{H}$ *by fractional linear transformations.*

PROOF: Let (i,ζ_0) be a fixed element of $S\mathcal{H}$, where ζ_0 is the unit vector at the point i tangent to the imaginary axis, and let (z,ζ) be an arbitrary element of $S\mathcal{H}$. There exists a unique $T\in PSL(2,\mathbb{R})$ sending the imaginary axis to the geodesic passing through z and tangent to ζ (Exercise 1.1) so that $T(i)=z$. By (2.1.1) we have $DT(\zeta_0)=\zeta$, and hence

$$T(i,\zeta_0)=(z,\zeta). \tag{2.1.2}$$

It is easy to see that the map $(z,\zeta) \to T$ is a homeomorphism between $S\mathcal{H}$ and $PSL(2,\mathbb{R})$.

For $S\in PSL(2,\mathbb{R})$, suppose that $S(z,\zeta)=(z',\zeta')$. By (2.1.2) $S(z,\zeta)=ST(i,\zeta_0)$, and hence $S(z,\zeta) \to ST$, and the last assertion follows. □

Let $d\ell=\sqrt{ds^2+d\theta^2}$ be a Riemannian metric on $S\mathcal{H}$, where ds is the hyperbolic metric on $\mathcal{H}$ (1.1.1), and $\theta=\frac{1}{2\pi}\arg(\zeta)$; and let $dv=d\mu d\theta$ be a volume on $S\mathcal{H}$, where $d\mu$ is the hyperbolic area on $\mathcal{H}$ (1.4.1). By Exercise 2.4, the metric $d\ell$ and the volume dv on $S\mathcal{H}$ are $PSL(2, \mathbb{R})$-invariant.

Besides being a group, $PSL(2, \mathbb{R})$ is also a topological space in which a transformation $z\to\frac{az+b}{cz+d}$ can be identified with the point $(a,b,c,d)\in\mathbb{R}^4$. More precisely, as a topological space, $SL(2, \mathbb{R})$ can be identified with the subset of $\mathbb{R}^4$,

$$X=\{(a,b,c,d)\in\mathbb{R}^4 \mid ad-bc=1\}.$$

If we define $\delta(a,b,c,d)=(-a,-b,-c,-d)$, then δ: $X\to X$ is a homeomorphism and δ together with the identity forms a cyclic group of order 2 acting on X. We topologize $PSL(2,\mathbb{R})$ as the quotient space. Exercise 2.1 shows that $PSL(2,\mathbb{R})$ is in fact a topological group. A *norm* on $PSL(2,\mathbb{R})$ is induced from $\mathbb{R}^4$: for $T(z)=\frac{az+b}{cz+d}$ with $ad-bc=1$, we define

$$\|T\| = (a^2+b^2+c^2+d^2)^{\frac{1}{2}}.$$

Notice that $\|T\|$ (as well as Tr(T), introduced in §1.1) is a well-defined function of T. PSL(2,$\mathbb{R}$) is a topological group with respect to the metric $\|T-S\|$. The group of all isometries of $\mathcal{H}$, Isom($\mathcal{H}$) is topologized similarly. Recall also that Isom($\mathcal{H}$) acts on $\mathcal{H}$ by homeomorphisms (Theorems 1.1.1 and 1.3.1.)

DEFINITION: A subgroup Γ of Isom($\mathcal{H}$) is called *discrete* if the induced topology on Γ is a discrete topology, i.e. if Γ is a discrete set in the topological space Isom($\mathcal{H}$).

As follows from Exercise 2.5, Γ is discrete if and only if $T_n \to \mathrm{Id}$, $T_n \in \Gamma$ implies $T_n = \mathrm{Id}$ for sufficiently large n.

2.2. Discrete and properly discontinuous groups

DEFINITION: A discrete subgroup of Isom($\mathcal{H}$) is called a *Fuchsian group* if it consists of orientation-preserving transformations, in other words, a Fuchsian group is a discrete subgroup of PSL(2,$\mathbb{R}$).

For any discrete group Γ of Isom($\mathcal{H}$), its subgroup Γ^+ of index ≤ 2 consisting of orientation-preserving transformations is a Fuchsian group. Thus the main ingredient in the study of discrete subgroups of isometries of $\mathcal{H}$ is the study of Fuchsian groups. The action of PSL(2,$\mathbb{R}$) on $\mathcal{H}$ lifts to the action on its unit tangent bundle S$\mathcal{H}$ by isometries (Exercise 2.4), thus sometimes it is useful to consider Fuchsian groups as discrete groups of isometries of S$\mathcal{H}$ (see §3.6). Discrete subgroups of Lie groups are sometimes called *lattices* by analogy with lattices in $\mathbb{R}^n$ which are discrete groups of isometries of

$\mathbb{R}^n$. The latter have the following important property: their action on $\mathbb{R}^n$ is *discontinuous* in the sense that every point of $\mathbb{R}^n$ has a neighborhood which is carried outside itself by all elements of the lattice except for the identity. In general, discrete groups of isometries do not have such discontinuous behavior, for if some elements have fixed points these points cannot have such a neighborhood. However, they satisfy a slightly weaker discontinuity condition. First we need several definitions.

Let X be a metric space, and let G be a group of homeomorphisms of X.

DEFINITION: A family $\{M_\alpha \mid \alpha \in A\}$ of subsets of X indexed by elements of a set A is called *locally finite* if for any compact subset $K \subset X$, $M_\alpha \cap K \neq \emptyset$ for only finitely many $\alpha \in A$.

REMARK: Some of the subsets M_α may coincide but they are still considered different elements of the family.

DEFINITION: For $x \in X$, a family $Gx = \{g(x) \mid g \in G\}$ is called the *G-orbit of the point* x. Each point of Gx is contained with a multiplicity equal to the order of G_x, the *stabilizer of* x *in* G.

DEFINITION: We say that a group G acts *properly discontinuously* on X if the G-orbit of any point $x \in X$ is locally finite.

It is clear from the definition that a group G acts properly discontinuously on X if and only if each orbit is discrete and the order of the stabilizer of each point is finite. In fact, the discreteness of all orbits already implies the discreteness of the group (see Corollary 2.2.7 for subgroups of $PSL(2,\mathbb{R})$). At least three more definitions of a properly discontinuous action equivalent to ours can be found in the literature [B,JS,VS]. (See Theorem 2.2.1 and Exercise 2.6.)

THEOREM 2.2.1. G *acts properly discontinuously on* X *if and only if each point* $x \in X$ *has a neighborhood* V *such that*

$$T(V) \cap V \neq \emptyset \text{ for only finitely many } T \in G. \tag{2.2.1}$$

PROOF: Suppose G acts properly discontinuously on X, then each orbit Gx is discrete, and for each point x, G_x is finite. This implies that for any point x there exists a ball $B_\epsilon(x)$ centered at x of radius ϵ containing no points of Gx other than x. Let $V \subset B_{\epsilon/2}(x)$ be a neighborhood of x, then $T(V) \cap V \neq \emptyset$ implies that $T \in G_x$, hence it is possible for only finitely many $T \in G$. Conversely, if (2.2.1) holds, we have to show that each G-orbit is discrete and that the stabilizer of each point z, G_z, has finite order. If Gz is not discrete, it has a limit point, say z_0, and any neighborhood of z_0 will meet infinitely many of its images under G, a contradiction with (2.2.1). Similarly, if $T(z)=z$ for infinitely many $T \in G$, then any neighborhood V of z meets infinitely many of its images under G. □

Before we give some examples of Fuchsian groups, let us identify discrete subgroups of one-dimensional Lie groups: $\mathbb{R}$, the additive group of real numbers, and S^1, the multiplicative group of complex numbers of modulus 1.

LEMMA 2.2.2.

(i) Any non-trivial discrete subgroup of $\mathbb{R}$ *is infinite cyclic.*

(ii) Any discrete subgroup of S^1 *is finite cyclic.*

PROOF of (i): Let Γ be a discrete subgroup of $\mathbb{R}$. Of course, $0 \in \Gamma$, and there exists a smallest positive $x \in \Gamma$, otherwise Γ would not be discrete. Then $\{nx \mid n \in \mathbb{Z}\}$ is a subgroup of Γ. Suppose there is $y \in \Gamma$, $y \neq nx$. We may assume $y>0$, otherwise we take $-y$ which also belongs to Γ. There exists an integer $k \geq 0$ such that $kx<y<(k+1)x$, and $y-kx<x$, and $(y-kx) \in \Gamma$ which contradicts the choice of x.

PROOF of (ii): Let Γ now be a discrete subgroup of $S^1=\{z \in \mathbb{C} \mid z=e^{i\phi}\}$. By discreteness there exists $z=e^{i\phi_0} \in \Gamma$, with the smallest argument ϕ_0, and for some $m \in \mathbb{Z}$, $m\phi_0=2\pi$, otherwise we get a contradiction with the choice of ϕ_0. □

According to the classification of elements of PSL(2,$\mathbb{R}$), one can form three types of cyclic subgroups of PSL(2,$\mathbb{R}$): hyperbolic, parabolic, and elliptic.

THEOREM 2.2.3.
(i) All hyperbolic and parabolic cyclic subgroups of PSL(2,$\mathbb{R}$) *are Fuchsian groups.*
(ii) An elliptic cyclic subgroup of PSL(2,$\mathbb{R}$) *is a Fuchsian group if and only if it is finite.*

We leave the proof of this theorem to the reader (Exercise 2.7).

EXAMPLE A. Let us consider a group which consists of all transformations

$$z \to \frac{az+b}{cz+d} \quad (a, b, c, d \in \mathbb{Z}, \ ad-bc=1).$$

It is called the *modular group* and denoted by PSL(2,$\mathbb{Z}$). It is clearly a discrete subgroup of PSL(2,$\mathbb{R}$) and hence a Fuchsian group. It will be studied in more detail in §§3.2, 3.5, and 5.5.

An extension of PSL(2,$\mathbb{Z}$), PS*L(2,$\mathbb{Z}$) is an example of a discrete group of isometries of $\mathcal{H}$ which is not a Fuchsian group.

Our next task is to show that an $\Gamma \subset$PSL(2, $\mathbb{R}$) is a Fuchsian group if and only if it acts properly discontinuously on $\mathcal{H}$.

LEMMA 2.2.4. *Let* $w_0 \in \mathcal{H}$ *be given and let* K *be a compact subset of* $\mathcal{H}$. *Then the set*

$$E=\{T\in PSL(2, \mathbb{R}) \mid T(z_0)\in K\}$$

is compact.

PROOF: PSL(2,$\mathbb{R}$) is topologized as a quotient space of SL(2,$\mathbb{R}$). Thus we have a continuous map ψ: SL(2,$\mathbb{R}$)$\to$PSL(2,$\mathbb{R}$) defined by

$$\psi\begin{bmatrix} a & b \\ c & d \end{bmatrix}=T, \quad \text{where } T(z)=\frac{az+b}{cz+d}.$$

If we show that

$$E_1=\left\{\begin{bmatrix} a & b \\ c & d \end{bmatrix}\in SL(2,\mathbb{R}) \mid \frac{az_0+b}{cz_0+d}\in K\right\}$$

is compact then it follows that $E=\psi(E_1)$ is compact. We prove that E_1 is compact by showing it is closed and bounded when regarded as a subset of $\mathbb{R}^4$ (identifying $\begin{bmatrix} a & b \\ c & d \end{bmatrix}$ with (a,b,c,d)). We have a continuous map β: SL(2,$\mathbb{R}$)$\to\mathcal{H}$ defined by $\beta(A)=\psi(A)(z_0)$. $E_1=\beta^{-1}(K)$, thus it follows that E_1 is closed as the inverse image of the closed set K.

We now show that E_1 is bounded. As K is bounded there exists $M_1>0$ such that

$$\left|\frac{az_0+b}{cz_0+d}\right|<M_1,$$

for all $\begin{bmatrix} a & b \\ c & d \end{bmatrix} \in E_1$.

Also, as K is compact in $\mathcal{H}$, there exists $M_2>0$ such that

$$\mathrm{Im}\left(\frac{az_0+b}{cz_0+d}\right)\geq M_2.$$

(1.1.3) implies that the left-hand side of this inequality is $\mathrm{Im}(z_0)/|cz_0+d|^2$ so that

$$|cz_0+d|\leq\sqrt{\left(\frac{\mathrm{Im}(z_0)}{M_2}\right)},$$

and thus

$$|az_0+b|\leq M_1\sqrt{\left(\frac{\mathrm{Im}(z_0)}{M_2}\right)},$$

and we deduce that a, b, c, d are bounded. □

LEMMA 2.2.5. *Let* Γ *be a subgroup of* PSL(2,$\mathbb{R}$) *acting properly discontinuously on* $\mathcal{H}$, *and* $p\in\mathcal{H}$ *be fixed by some element of* Γ. *Then there is a neighborhood* W *of* p *such that no other point of* W *is fixed by an element of* Γ *other than the identity.*

PROOF: Suppose $T(p)=p$ for some $\mathrm{Id}\neq T\in\Gamma$, and in any neighborhood of p there are fixed points of transformations in Γ, i.e. there is a sequence of points in $\mathcal{H}$, $p_n\to p$, such that for $T_n\in\Gamma$, $T_n(p_n)=p_n$. Let $\overline{B_{3\epsilon}(p)}$ be a closed hyperbolic disc, centered at p, of radius $3\epsilon>0$. As the topology induced by the hyperbolic metric coincides with the Euclidean topology (Theorem 1.3.3), $\overline{B_\epsilon(p)}$ is compact. Since Γ acts properly discontinuously, the set $\left\{T\in\Gamma \mid T(p)\in\overline{B_{3\epsilon}(p)}\right\}$ is finite. Hence, for N sufficiently large, $n>N$ implies that $\rho(T_n(p), p)>3\epsilon$ while $\rho(p_n, p)<\epsilon$. By the triangle inequality and the invariance of the hyperbolic metric (Theorem

1.1.2) we have

$$\rho(T_n(p), p) \le \rho(T_n(p), T_n(p_n)) + \rho(T_n(p_n), p) = \rho(p, p_n) + \rho(p_n, p) < 2\epsilon,$$

a contradiction. □

THEOREM 2.2.6. *Let* Γ *be a subgroup of* PSL(2,$\mathbb{R}$). *Then* Γ *is a Fuchsian group if and only if* Γ *acts properly discontinuously on* $\mathcal{H}$.

PROOF: We first show that a Fuchsian group acts properly discontinuously on $\mathcal{H}$. Let $z \in \mathcal{H}$ and K be a compact subset of $\mathcal{H}$. Then $\{T \in \Gamma \mid T(z) \in K\} = \{T \in PSL(2,\mathbb{R}) \mid T(z) \in K\} \cap \Gamma$ is a finite set (it is the intersection of a compact and a discrete set), and hence Γ acts properly discontinuously. Conversely, suppose Γ acts properly discontinuously, but it is not a discrete subgroup of PSL(2,$\mathbb{R}$). Choose a point $s \in \mathcal{H}$ not fixed by any non-identity element of Γ: such points exist by Lemma 2.2.5. As we are assuming that Γ is not discrete, there exists a sequence $\{T_k\}$ of distinct elements of Γ such that $T_k \to Id$ as $k \to \infty$. Hence $T_k(s) \to s$ as $k \to \infty$ and as s is not fixed by any non-identity element of Γ, $\{T_k(s)\}$ is a sequence of points distinct from s. Hence every closed hyperbolic disc centered at s contains infinitely many points of the Γ-orbit of s, and hence Γ does not act properly discontinuously. □

COROLLARY 2.2.7. *Let* Γ *be a subgroup of* PSL(2, $\mathbb{R}$). *Then* Γ *acts properly discontinuously on* $\mathcal{H}$ *if and only if for all* $z \in \mathcal{H}$, Γz, *the* Γ*-orbit of* z, *is a discrete subset of* $\mathcal{H}$.

PROOF: Suppose Γ acts properly discontinuously on $\mathcal{H}$, hence each Γ-orbit is a locally finite family of points, hence a discrete set of $\mathcal{H}$. Conversely, suppose Γ does not act properly discontinuously on $\mathcal{H}$ and hence by Theorem 2.2.6 is not discrete. Repeating the argument in the proof of Theorem 2.2.6, we construct a sequence $\{T_k(s)\}$ of points distinct from s such that $T_k(s) \to s$, hence the Γ-orbit of the point s is not discrete. □

COROLLARY 2.2.8. *If Γ is a Fuchsian group, then the fixed points of elliptic elements do not accumulate in $\mathcal{H}$.*

PROOF: Let $z \in \mathcal{H}$, and K be a compact set containing z. Suppose $z = Tz$ for some $T \in \Gamma$, then $K \cap T(K) \neq \emptyset$. But Exercise 2.6 and Theorem 2.2.6 imply that this is only possible for a finite number of $T \in \Gamma$, hence there are only finitely many elliptic fixed points in K, and the Corollary follows. □

Exercise 2.9 shows that a properly discontinuous action of a subgroup of PSL(2,$\mathbb{R}$) lifts to the unit tangent bundle $S\mathcal{H}$. However, discreteness of a group does not always imply its discontinuous action. For example, the modular group does not act properly discontinuously on $\mathbb{R} \cup \{\infty\}$: the orbit of 0 is the set $\mathbb{Q} \cup \{\infty\}$ ($\mathbb{Q}$ is the set of rational numbers) which is dense in $\mathbb{R} \cup \{\infty\}$.

Corollary 2.2.7 implies the following: if $z \in \mathcal{H}$ and $\{T_n\}$ is a sequence of distinct elements in Γ, then if $\{T_n(z)\}$ has a limit point $\alpha \in \mathbb{C} \cup \{\infty\}$ then $\alpha \in \mathbb{R} \cup \{\infty\}$.

DEFINITION. The set of all possible limit points of Γ-orbits Γz, $z \in \mathcal{H}$ is called the *limit set* of Γ and denoted by $\Lambda(\Gamma)$.

Thus for all Fuchsian groups Γ, $\Lambda(\Gamma) \subseteq \mathbb{R} \cup \{\infty\}$.

EXAMPLE B. If Γ is the cyclic group generated by $z \to 2z$, then $\Lambda(\Gamma) = \{0, \infty\}$.

EXAMPLE A. If Γ is the modular group (see §2.2, Example A), then $\Lambda(\Gamma)=\mathbb{R}\cup\{\infty\}$ (this will follow from Theorem 4.5.2).

2.3. Algebraic properties of Fuchsian groups

If G is any group and $g\in G$, then the *centralizer* of g in G is defined by

$$C_G(g)=\{h\in G \mid hg=gh\}.$$

LEMMA 2.3.1. *If* $ST=TS$ *then* S *maps the fixed-point set of* T *to itself.*

PROOF: Suppose that T fixes p. Then

$$S(p)=ST(p)=TS(p),$$

so that S(p) is also fixed by T. □

Let us look at centralizers of parabolic, elliptic, and hyperbolic elements in $PSL(2,\mathbb{R})$. Suppose that $T(z)=z+1$. If $S\in C_{PSL(2,R)}(T)$ then $S(\infty)=\infty$. Therefore, $S(z)=az+b$. $ST=TS$ gives us $a=1$. Hence

$$C_{PSL(2,R)}(T)=\{z\to z+k \mid k\in\mathbb{R}\}.$$

The centralizer of an elliptic transformation of the unit disc $\mathcal{U}$ fixing 0 (i.e. $z\to e^{i\phi}z$) consists of all transformations of the form $z\to\frac{az+\bar{c}}{cz+\bar{a}}$ fixing 0, i.e of the form $z\to e^{i\theta}z$ $(0\le\theta<2\pi)$. Let $T(z)=\lambda z$ $(\lambda>0,\ \lambda\ne1)$ and $S\in C_{PSL(2,R)}(T)$. Then a direct calculation shows that S is given by a diagonal matrix and hence $S(z)=\mu z$ $(\mu>0)$. From these calculations we deduce the following results.

THEOREM 2.3.2. *Two non-identity elements of* PSL(2,ℝ) *commute if and only if they have the same fixed-point set.*

THEOREM 2.3.3. *The centralizer in* PSL(2,ℝ) *of a hyperbolic (resp. parabolic, elliptic) element of* PSL(2,ℝ) *consists of all hyperbolic (resp. parabolic, elliptic) elements with the same fixed-point set, together with the identity element.*

COROLLARY 2.3.4. *Two hyperbolic elements in* PSL(2,ℝ) *commute if and only if they have the same axes.*

In what follows, we are going to use the following important property of the trace function: $\mathrm{Tr}(S^{-1}TS)=\mathrm{Tr}(T)$ for any T, S∈PSL(2,ℝ). This implies that the type of an element in PSL(2,ℝ) is invariant under *conjugation.*

THEOREM 2.3.5. *Let* Γ *be a Fuchsian group all of whose non-identity elements have the same fixed-point set. Then* Γ *is cyclic.*

PPOOF: We recall first that the fixed-point set of an element of PSL(2,ℝ) (two points in $\mathbb{R}\cup\{\infty\}$, one point in $\mathbb{R}\cup\{\infty\}$, or one point in $\mathcal{H}$) defines its type, hence all elements of Γ must be of the same type. Suppose all elements of Γ are hyperbolic. Then by choosing a conjugate group we may assume that each S∈Γ fixes 0 and ∞. Thus Γ is a discrete subgroup of $H=\{z\to\lambda z \mid \lambda>0\}$ which is isomorphic as a topological group to $\mathbb{R}^*$, the multiplicative group of positive real numbers. $\mathbb{R}^*$ is isomorphic as a topological group to $\mathbb{R}$ via the isomorphism $x\to\ln x$. Hence by Lemma 2.2.2(i), Γ is infinite cyclic. Similarly, if Γ contains a parabolic element, then Γ is an infinite cyclic group containing only parabolic elements. Suppose Γ contains an elliptic element. In $\mathcal{U}$, the unit disc model, Γ is a discrete subgroup of orientation-preserving isometries of $\mathcal{U}$. By choosing a conjugate group we may assume that all elements of Γ have 0 as a fixed point, and therefore all elements of Γ are of the form $z \to e^{i\phi}z$. Thus Γ is isomorphic to a subgroup of S^1, and it is discrete if and only if the corresponding subgroup of S^1 is discrete. Now the assertion follows from Lemma 2.2.2(ii). □

THEOREM 2.3.6. *Every abelian Fuchsian group is cyclic.*

PROOF: By Theorem 2.3.2, all non-identity elements in an abelian Fuchsian group have the same fixed-point set. The theorem follows now immediately from Theorem 2.3.5. □

COROLLARY 2.3.7. *No Fuchsian group is isomorphic to* $\mathbb{Z}\times\mathbb{Z}$.

If G is a group and H is a subgroup of G, then the *normalizer* $N_G(H)$ of H in G is

$$N_G(H)=\{g\in G \mid gHg^{-1}=H\}.$$

THEOREM 2.3.8. *Let* Γ *be a non-abelian Fuchsian group. Then the normalizer of* Γ *in* $PSL(2,\mathbb{R})$ *is a Fuchsian group.*

PROOF: Suppose that the normalizer of Γ in $PSL(2,\mathbb{R})$ is not Fuchsian. Then it contains an infinite sequence $\{T_i\}$ of distinct elements such that $T_i \to Id$ as $i\to\infty$. Thus if $S\in\Gamma$ $(S\neq Id)$, then $T_iST_i^{-1}\to S$ as $i\to\infty$. Since $T_iST_i^{-1}\in\Gamma$ and Γ is discrete, there exists an integer m such that $T_iST_i^{-1}=S$ for $i>m$. Thus for these values of i, Theorem 2.3.2 implies that T_i has the same fixed-point set as S. Now as Γ is not abelian, Theorem 2.3.2 implies that there exists $S'\in\Gamma$ with a different fixed-point set from that of S. However, by the same argument T_i has the same fixed-point set as S' for sufficiently large i and hence S' has the same fixed-point set as S, a contradiction. □

2.4. Elementary groups

It is easily follows from (1.1.3) that the action of the group PSL(2,$\mathbb{R}$) extends from $\mathcal{H}$ to its *Euclidean boundary* $\mathbb{R}\bigcup\{\infty\}$, hence PSL(2,$\mathbb{R}$) acts on the *Euclidean closure* of $\mathcal{H}$, denoted by $\widetilde{\mathcal{H}}$ (see §1.2).

DEFINITION. A subgroup Γ of PSL(2,$\mathbb{R}$) is called *elementary* if there exists a finite Γ-orbit in $\widetilde{\mathcal{H}}$.

Since $\mathcal{H}$ and $\mathbb{R}\cup\{\infty\}$ are PSL(2,$\mathbb{R}$)-invariant, any Γ-orbit of a point in $\widetilde{\mathcal{H}}$ is either all in $\mathcal{H}$ or all in $\mathbb{R}\cup\{\infty\}$.

Let $g,h\in PSL(2,\mathbb{R})$ and $[g,h]=g\circ h\circ g^{-1}\circ h^{-1}\in\Gamma$ be the *commutator* of g and h. It is useful to notice that tr[g,h] does not depend on the choice of the matrices, representing g and h, hence is a well-defined function of g and h.

THEOREM 2.4.1. *Let* Γ *be a subgroup of* PSL(2,$\mathbb{R}$) *containing besides the identity only elliptic elements. Then all elements of* Γ *have the same fixed point, and hence* Γ *is a cyclic group, abelian and elementary.*

PROOF: We shall prove that all elliptic elements in Γ must have the same fixed point. In the unit disc model, let us conjugate Γ in such a way that an element $Id\neq g\in\Gamma$ fixes 0: $g=\begin{bmatrix} u & 0 \\ 0 & \bar{u} \end{bmatrix}$, and let $h=\begin{bmatrix} a & \bar{c} \\ c & \bar{a} \end{bmatrix}\in\Gamma$, $h\neq g$. We have $tr[g,h]=2+4|c|^2(\mathrm{Im}(u))^2$. Since Γ does not contain hyperbolic elements, $|tr[g,h]|\leq 2$, and either $\mathrm{Im}(u)=0$

or c=0. If Im(u)=0 then $u=\bar{u}\in\mathbb{R}$ and hence g=Id, a contradiction. Hence c=0, and so $h=\begin{bmatrix} a & 0 \\ 0 & \bar{a} \end{bmatrix}$ also fixes 0. Thus we conclude that all elements of Γ have the same fixed point. So by Theorem 2.3.5, Γ is a finite cyclic group, and hence abelian. Since 0 is a Γ-orbit, Γ is elementary. □

COROLLARY 2.4.2. *Any Fuchsian group containing besides the identity only elliptic elements is a finite cyclic group.*

The next theorem describes all elementary Fuchsian groups.

THEOREM 2.4.3. *Any elementary Fuchsian group is either cyclic or is conjugate in* PSL(2,$\mathbb{R}$) *to a group generated by* $g(z)=kz$ $(k>1)$ *and* $h(z)=-1/z$.

PROOF: *Case 1.* Suppose Γ fixes a single point $\alpha\in\widetilde{\mathcal{H}}$. If $\alpha\in\mathcal{H}$, then all elements of Γ are elliptic; and by Corollary 2.4.2, Γ is a finite cyclic group. Suppose $\alpha\in\mathbb{R}\cup\{\infty\}$. Then Γ cannot have elliptic elements. We shall show that hyperbolic and parabolic elements cannot have a common fixed point. Assume the opposite, and suppose this point is ∞, and $g(z)=\lambda z$ $(\lambda>1)$ and $h(z)=z+k$ (since g and h have only one common fixed point, $k\neq0$). Then $g^{-n}\circ h\circ g^{n}(z)=z+\lambda^{-n}k$. Since $\lambda>1$, we find that the sequence

$$\|g^{-n}\circ h\circ g^{n}\| \quad (\ n=1,2,\ldots\ )$$

is bounded; hence $\{g^{-n}\circ h\circ g^{n}\}$ contains a convergent subsequence of distinct terms which contradicts the discreteness of Γ. We conclude that Γ can only contain elements of one type. If Γ contains only parabolic elements, by Theorem 2.3.5 it is an infinite cyclic group. Suppose Γ contains only hyperbolic elements. We shall prove that in this case their second fixed points must also coincide, and so Γ will fix two points in $\mathbb{R}\cup\{\infty\}$. Suppose $f(z)=\lambda^2 z$ ($\lambda>1$) (it fixes 0 and ∞) and $g(z)=\frac{az+b}{cz+d}$ which fixes 0 but not ∞. Then $b=0$, $c\neq 0$, $a\neq 0$, and $d=1/a$. Then $f\circ g\circ f^{-1}\circ g^{-1}$ is given by the matrix $\begin{bmatrix} 1 & 0 \\ t & 1 \end{bmatrix}$ with $t=\frac{c}{a}\left(\frac{1}{\lambda^2}-1\right)$. Since $c\neq 0$ we obtain a parabolic element in Γ, a contradiction.

Case 2. Now suppose Γ has an orbit in $\mathbb{R}\cup\{\infty\}$ consisting of two points. An element of Γ either fixes each of them or interchanges them. A parabolic element cannot fix two points. Since each orbit (except for a single fixed point of a parabolic transformation) is infinite, a parabolic element cannot interchange these points; hence Γ does not contain any parabolic elements. All hyperbolic elements must have the same fixed point set. If Γ contains only hyperbolic elements, then it is cyclic by Theorem 2.3.6. If it contains only elliptic elements, it is finite cyclic by Corollary 2.4.2. If Γ contains both hyperbolic and elliptic elements, it must contain an elliptic element of order 2 interchanging the common fixed points of the hyperbolic elements; and then Γ is conjugate to a group generated by $g(z)=kz$ ($k>1$) and $h(z)=-1/z$.

Case 3. Suppose now Γ has an orbit in $\mathcal{H}$ consisting of $k=2$ points or an orbit in $\widetilde{\mathcal{H}}$ consisting of $k\geq 3$ points. Since the parabolic and hyperbolic elements can have only either fixed points at infinity or infinite orbits, Γ must contain only elliptic elements and therefore is a finite cyclic group; and it is conjugate to a group generated by $z\to e^{\frac{2\pi i}{k}}z$. □

THEOREM 2.4.4. *A non-elementary subgroup* Γ *of* PSL(2,$\mathbb{R}$) *must contain a hyperbolic element.*

PROOF: Suppose Γ does not contain hyperbolic elements. If Γ contains only elliptic elements (and Id), then by Theorem 2.4.1 it is elementary. Hence Γ contains a parabolic element fixing, say, ∞: $f(z)=z+1$. Let $g(z)=\frac{az+b}{cz+d}$ be any element in Γ. Then $f^n \circ g(z)=\frac{(a+nc)z+(b+nd)}{cz+d}$. We have, thus,

$$\mathrm{tr}^2(f^n \circ g)=(a+d+nc)^2.$$

Since all elements in the group are either elliptic or parabolic, we have $0 \leq (a+d+nc)^2 \leq 4$ for all n; which implies $c=0$. But then g also fixes ∞, so that ∞ is fixed by all elements in Γ; hence Γ is elementary, a contradiction. □

In fact, any non-elementary subgroup of $PSL(2,\mathbb{R})$ must contain infinitely many hyperbolic elements, no two of which have a common fixed point (see Exercise 2.13).

THEOREM 2.4.5. *If Γ, a subgroup $PSL(2,\mathbb{R})$, contains no elliptic elements it is either elementary or discrete.*

PROOF: Assume Γ is non-elementary. Then by Theorem 2.4.4 it contains a hyperbolic element h. We may assume that h is given by the matrix $\begin{bmatrix} u & 0 \\ 0 & 1/u \end{bmatrix}$, $u>0$. In order to prove that Γ is discrete we must show that for any sequence $g_n \to \mathrm{Id}$ $(g_n \in \Gamma)$, $g_n=\mathrm{Id}$ for sufficiently large n.

Let $g_n=\begin{bmatrix} a_n & b_n \\ c_n & d_n \end{bmatrix}$, $a_nd_n-b_nc_n=1$, be such a sequence. An easy calculation shows that, since $g_n \to \mathrm{Id}$, $\mathrm{tr}(h \circ g_n \circ h^{-1} \circ g_n^{-1})$ $=(2-b_nc_n(u-\frac{1}{u})^2) \to 2$ as $n \to \infty$. Since Γ contains no elliptic elements, $|\mathrm{tr}[h,g_n]|\geq 2$, so we conclude that for sufficiently large n

$$b_n c_n \leq 0.$$

Let $f_n=[h,g_n]=h\circ g_n\circ h^{-1}\circ g_n^{-1}$ be given by a matrix $\begin{bmatrix} A_n & B_n \\ C_n & D_n \end{bmatrix}$, $A_nD_n-B_nC_n=1$. As $n\to\infty$, we have $f_n\to \mathrm{Id}$, since $g_n\to \mathrm{Id}$. Therefore by the same argument as above, we obtain for sufficiently large n

$$B_n C_n \leq 0.$$

On the other hand, $\mathrm{tr}[h,f_n]=2-B_nC_n(u-\frac{1}{u})^2$ $=2+b_nc_n(1+b_nc_n)(u-\frac{1}{u})^4\to 2$ as $n\to\infty$, and from this we obtain for sufficiently large n

$$b_n c_n \geq 0.$$

Finally, we conclude that there exists $N>0$ such that for $n>N$ $b_nc_n=0$. Hence for $n>N$, h and g_n have a common fixed point: 0 if $b_n=0$, and ∞ if $c_n=0$.

To complete the proof, we apply Exercise 2.13 and construct three hyperbolic elements h_1, h_2, and h_3 in Γ no two of which have a common fixed point. Then for $n>N$, g_n has a common fixed point with each of h_1, h_2 and h_3, no two of which coincide. Therefore g_n has three fixed points, and thus $g_n=\mathrm{Id}$. □

The following theorem shows that two transformations generating a discrete non-elementary group cannot approach too closely to the identity.

Let $\langle T, S\rangle$ denote a group generated by transformations T and S.

THEOREM 2.4.6. (Jørgensen inequality) *Suppose that* $T,S\in PSL(2,\mathbb{R})$ *and* $\langle T, S\rangle$ *is a discrete non-elementary group. Then*

$$|tr^2(T)-4|+|tr(TST^{-1}S^{-1})-2|\geq 1. \tag{2.5.1}$$

The lower bound is best possible.

First we prove the following lemma:

LEMMA 2.4.7. *Suppose* T, S∈PSL(2, ℝ) *and* T≠Id. *Define* $S_0=S$, $S_1=S_0\circ T\circ S_0^{-1}$, ... , $S_{r+1}=S_r\circ T\circ S_r^{-1}$, ... *If, for some* n, $S_n=T$, *then* <T,S> *is elementary and* $S_2=T$.

PROOF: Suppose first that T has one fixed point α (i.e. T is parabolic or elliptic). Notice that the S_r ($r\geq 1$) are conjugate to T and hence also have one fixed point. We have $S_{r+1}\circ S_r(\alpha)=S_r\circ T\circ S_r^{-1}\circ S_r(\alpha)=S_r(\alpha)$, i.e. S_{r+1} fixes $S_r(\alpha)$. If S_{r+1} fixes α, so does S_r. As $S_n(=T)$ fixes α, we conclude that each S_j (including $S_0=S$) fixes α. So all elements in <T,S> fix α. If T is parabolic, we conclude that <T,S> is elementary. If T is elliptic, it follows that all elements in <T,S> are elliptic; hence <T,S> is finite cyclic and therefore elementary. Also S_1 fixes α and therefore commutes with T, so $S_2=T$.

Suppose now that T has exactly two fixed points; we may assume then that $T(z)=kz$. Clearly S_1, ... , S_n have exactly two fixed points. The same argument as above shows that for $0\leq r\leq n$

$$\{0,\infty\}=\{S_r(0),S_r(\infty)\}.$$

But for $r\geq 1$, S_r is conjugate to T, and therefore it cannot interchange two points (all orbits of a hyperbolic transformation are infinite with the exception of two fixed points). Hence S_1, ... ,S_n fix 0 and ∞, and both $S=S_0$ and T leave the set $\{0,\infty\}$ invariant. Therefore <T,S> is elementary. □

PROOF of THEOREM: If T is of order 2 then tr(T)=0, and (2.5.1) holds. Suppose now T is not of order 2. Define S_0, S_1, ... as in Lemma 2.4.7. We are going to show that if (2.5.1) fails, then for some n we have

$$S_n = T, \tag{2.5.2}$$

which, by Lemma 2.4.7 will imply that $<T,S>$ is elementary, a contradiction. Since the expression in (2.5.1) is well-defined for $T,S \in PSL(2,\mathbb{R})$, we shall work with matrices representing these transformations.

Case 1. T is parabolic. Since the trace is invariant under conjugation, we may assume that

$$T = \begin{bmatrix} 1 & 1 \\ 0 & 1 \end{bmatrix}, \quad S = \begin{bmatrix} a & b \\ c & d \end{bmatrix}$$

where $c \neq 0$ (else $<T,S>$ is elementary). We are assuming that (2.5.1) fails, i.e. that $|c|<1$ (Exercise 2.12.) Writing $S_n = \begin{bmatrix} a_n & b_n \\ c_n & d_n \end{bmatrix}$ we obtain from

$$S_{n+1} = S_n \circ T \circ S_n^{-1} \tag{2.5.3}$$

$$\begin{bmatrix} a_{n+1} & b_{n+1} \\ c_{n+1} & d_{n+1} \end{bmatrix} = \begin{bmatrix} a_n & b_n \\ c_n & d_n \end{bmatrix} \begin{bmatrix} 1 & 1 \\ 0 & 1 \end{bmatrix} \begin{bmatrix} d_n & -b_n \\ -c_n & a_n \end{bmatrix} =$$

$$\begin{bmatrix} 1-a_n c_n & a_n^2 \\ -c_n^2 & 1+a_n c_n \end{bmatrix}.$$

By induction we deduce that $c_n = -(-c)^{2^n}$ which is equal to $-c^{2^n}$ for $n>0$, and as $|c|<1$

$$c_n \to 0.$$

Since $|c_n|<1$, we have by induction that $|a_n| \leq n+|a|$; so $a_n c_n \to 0$, and hence

$$a_{n+1} \to 1.$$

Thus $S_{n+1} \to T$, which by discreteness yields (2.5.2) for large n.

Case 2. T is hyperbolic. We may assume that

$$T = \begin{bmatrix} u & 0 \\ 0 & 1/u \end{bmatrix} \quad (u>1), \quad S = \begin{bmatrix} a & b \\ c & d \end{bmatrix},$$

and $bc \neq 0$ (else $<T,S>$ is elementary). If (2.5.1) fails, then

$$\mu = |\mathrm{tr}^2(T) - 4| + |\mathrm{tr}(TST^{-1}S^{-1}) - 2| = (1+|bc|)|u - \tfrac{1}{u}|^2 < 1.$$

Rewriting (2.5.3) we obtain

$$\begin{bmatrix} a_{n+1} b_{n+1} \\ c_{n+1} d_{n+1} \end{bmatrix} = \begin{bmatrix} a_n d_n u - \frac{b_n c_n}{u} & a_n b_n(\frac{1}{u} - u) \\ c_n d_n(u - \frac{1}{u}) & \frac{a_n d_n}{u} - b_n c_n u \end{bmatrix},$$

so $b_{n+1} c_{n+1} = -b_n c_n (1 + b_n c_n)(u - \frac{1}{u})^2$. By induction

$$|b_n c_n| \leq \mu^n |bc| \leq |bc|.$$

Therefore $b_n c_n \to 0$, $a_n d_n = 1 + b_n c_n \to 1$, $a_{n+1} \to u$, $d_{n+1} \to 1/u$.

$$\left|\frac{b_{n+1}}{b_n}\right| = |a_n(\tfrac{1}{u} - u)| \to |u(\tfrac{1}{u} - u)| \leq \mu^{\frac{1}{2}} |u|.$$

So $\left|\frac{b_{n+1}}{u^{n+1}}\right| < \mu^{\frac{1}{2}} \left|\frac{b_n}{u^n}\right|$ for sufficiently large n. Hence $\frac{b_n}{u^n} \to 0$,

and similarly,

$$c_n u^n \to 0.$$

Therefore

$$T^{-n} S_{2n} T^n = \begin{bmatrix} a_{2n} & b_{2n}/u^{2n} \\ c_{2n} u^{2n} & d_{2n} \end{bmatrix} \to T.$$

Since $<T,S>$ is discrete, for large n we have

$$T^{-n} S_{2n} T^n = T,$$

and hence $S_{2n} = T$ which is (2.5.2).

Case 3. T is elliptic. Using the unit disc model we may assume that

$$T=\begin{bmatrix} u & 0 \\ 0 & 1/u \end{bmatrix}$$

where $u \in \mathbb{C}$, $|u|=1$, and the same proof as in Case 2 works.

Now we show that the lower bound (2.5.1) is the best possible. Consider the group generated by $T(z)=z+1$ and $S(z)=-1/z$. As we will see in §3.1, $<T,S>=PSL(2,\mathbb{Z})$ which is discrete and non-elementary. We have $T \circ S \circ T^{-1} \circ S^{-1}(z)=\frac{2z+1}{z+1}$ with trace 3, and hence the equality holds in (2.5.1). □

The following theorem gives a general criterion for discreteness.

THEOREM 2.4.8. *A non-elementary subgroup* Γ *of* $PSL(2, \mathbb{R})$ *is discrete if and only if, for each* T *and* S *in* Γ, *the group* $<T,S>$ *is discrete.*

PROOF: If Γ is discrete, then every subgroup of it is also discrete. Suppose now that every subgroup $<T, S>$ is discrete, but Γ itself is not. Then we can find a sequence of distinct transformations in Γ, $T_1, T_2, \ldots, T_n, \ldots$ such that $T_n \neq Id$, and $\lim_{n\to\infty} T_n = Id$. Since $T^2=Id$ implies $Tr(T)=0$ (an easy calculation), and $Tr(T)$ is a continuous function on $PSL(2, \mathbb{R})$, we may choose a subsequence that contains no elements of order 2. For any $S \in \Gamma$ we have

$$|tr^2(T_n)-4|+|tr(T_nST_n^{-1}S^{-1})-2| \to 0;$$

and so by Theorem 2.4.6, for $n \geq n(S)$ say, the group $<T_n, S>$ is elementary. By Exercise 2.13, Γ contains two hyperbolic elements S_1 and S_2 with no common fixed points. For $n \geq \max(n(S_1), n(S_2))$, both groups

$$<T_n, S_1> \text{ and } <T_n, S_2>$$

are elementary and discrete and according to Theorem 2.4.3, T_n must leave the fixed point pair of S_1 and that of S_2 invariant. As T_n is not elliptic of order 2, it cannot interchange a pair of points; so T_n must fix each individual fixed point of S_1 and of S_2. Since their fixed points do not coincide, T_n must fix four distinct points which implies $T_n = Id$, a contradiction. □

EXERCISES FOR CHAPTER 2

2.1. Prove that group multiplication and taking of inverses are continuous with respect to the topology on $PSL(2,\mathbb{R})$.

2.2. Prove that $\|T\|^2=2\cosh \rho(i,T(i))$.

2.3. Show that every transformation in $PSL(2,\mathbb{R})$ can be written uniquely in the form TR, where R is an elliptic element fixing i and $T(z)=az+b$ ($a,b\in\mathbb{R}$, $a>0$). Deduce that as a topological space $PSL(2,\mathbb{R})$ is homeomorphic to $\mathbb{R}^2\times S^1$, where S^1 is a circle.

2.4. Prove that the metric $d\ell$ and the volume dv on $S\mathcal{H}$ are $PSL(2,\mathbb{R})$-invariant.

2.5. Prove that $\Gamma\subset Isom(\mathcal{H})$ is discrete if and only if $T_n\to Id$ ($T_n\in\Gamma$) implies $T_n=Id$ for sufficiently large n.

2.6. Let X and G be as in § 2.2. Prove that the following statements are equivalent:

(i) G acts properly discontinuously on X;

(ii) For any compact set K in X, $T(K)\cap K\neq\emptyset$ for only finitely many $T\in G$;

(iii) Any point $x\in X$ has a neighborhood V such that $T(V)\cap V\neq\emptyset$ implies $T(x)=x$.

2.7. Prove Theorem 2.2.3.

2.8. Prove that any Fuchsian group is countable.

2.9. Prove that if Γ is a Fuchsian group then it acts properly discontinuously on $S\mathcal{H}$.

2.10. Prove that the group

$$\Gamma=\{z\to\frac{(a_1+b_1\sqrt{2})z+(a_2+b_2\sqrt{2})}{(a_3+b_3\sqrt{2})z+(a_4+b_4\sqrt{2})} \mid$$

$$a_i,\ b_i\in\mathbb{Z},\ (a_1+b_1\sqrt{2})(a_4+b_4\sqrt{2})-(a_2+b_2\sqrt{2})(a_3+b_3\sqrt{2})=1\}$$

is a subgroup of $PSL(2,\mathbb{R})$ but is not a Fuchsian group.

2.11. Prove that a Fuchsian group Γ is elementary if and only if for any T, S$\in\Gamma$, the subgroup <T, S> is elementary.

2.12. Let $T=\begin{bmatrix} 1 & 1 \\ 0 & 1 \end{bmatrix}$, $S=\begin{bmatrix} a & b \\ c & d \end{bmatrix}$ be two matrices in SL(2,$\mathbb{R}$). Prove that the Jørgensen inequality (2.5.1) for T and S holds if and only if $|c|\geq 1$.

2.13. Prove that any non-elementary subgroup of PSL(2,$\mathbb{R}$) must contain infinitely many hyperbolic elements, no two of which have a common fixed point.

2.14. Let Γ be a non-elementary subgroup of PSL(2,$\mathbb{R}$). Then the following are equivalent:

(i) Γ is discrete;

(ii) The fixed points of elliptic elements do not accumulate in $\mathcal{H}$;

(iii) The elliptic elements do not accumulate to Id;

(iv) Each elliptic element has finite order.

3. FUNDAMENTAL REGIONS

3.1. Definition of a fundamental region

We are going to be concerned with fundamental regions of mainly Fuchsian groups, however it is convenient to give a definition in a slightly more general situation. As in §2.2, let X be a metric space, and G be a group of homeomorphisms acting properly discontinuously on X.

DEFINITIONS. A closed region $F \subset X$ (i.e. a closure of a non-empty open set $\mathring{F}$, called the interior of F) is defined to be a *fundamental region* for G if

(i) $\bigcup_{T \in G} T(F) = X$,

(ii) $\mathring{F} \cap T(\mathring{F}) = \emptyset$ for all $T \in G - \{Id\}$.

The set $\partial F = F - \mathring{F}$ is called the *boundary* of F.

The family $\{T(F) \mid T \in G\}$ is called the *tessellation* of X.

We shall prove in §3.2 that any Fuchsian group possesses a nice (connected and convex) fundamental region. Now we give an example in the simplest situation.

EXAMPLE B. Let Γ be the cyclic group generated by the transformation $z \to 2z$. Then the semi-annulus shown in Figure 7a is easily seen to be a fundamental region for Γ. It is already clear from this example that a fundamental region is not uniquely determined by the group: an arbitrary small perturbation of the lower semicircle determines a perturbation of the upper semicircle, and gives yet another fundamental region shown in Figure 7b.

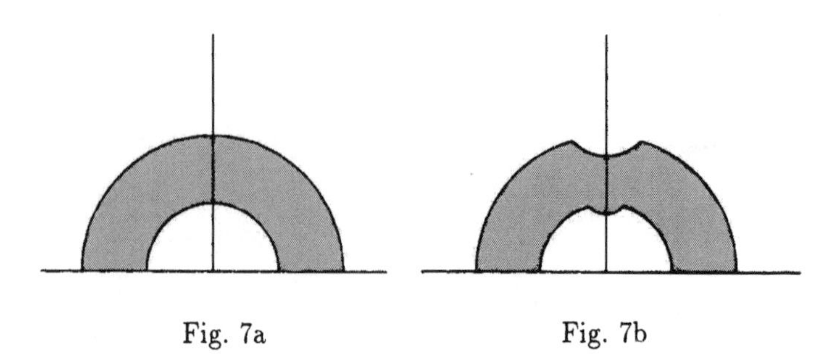

Fig. 7a Fig. 7b

THEOREM 3.1.1. *Let* F_1 *and* F_2 *be two fundamental regions for a Fuchsian group* Γ, *and* $\mu(F_1)<\infty$. *Suppose that the boundaries of* F_1 *and* F_2 *have zero hyperbolic area. Then* $\mu(F_2)=\mu(F_1)$.

PROOF: We have $\mu(\mathring{F}_i)=\mu(F_i)$, i=1,2. Now

$$F_1 \supseteq F_1 \cap \left(\bigcup_{T\in\Gamma} T(\mathring{F}_2)\right)=\bigcup_{T\in\Gamma} (F_1 \cap T(\mathring{F}_2)).$$

Since $\mathring{F}_2$ is the interior of a fundamental region, the sets $F_1 \cap T(\mathring{F}_2)$ are disjoint, and hence

$$\mu(F_1)\geq\sum_{T\in\Gamma} \mu(F_1\cap T(\mathring{F}_2))=\sum_{T\in\Gamma} \mu(T^{-1}(F_1)\cap\mathring{F}_2)=\sum_{T\in\Gamma} \mu(T(F_1)\cap\mathring{F}_2).$$

Since F_1 is a fundamental region

$$\bigcup_{T\in\Gamma} T(F_1)=\mathcal{H},$$

and therefore

$$\bigcup_{T\in\Gamma} (T(F_1)\cap\mathring{F}_2)=\mathring{F}_2.$$

Hence

$$\sum_{T\in\Gamma} \mu(T(F_1)\cap\mathring{F}_2)\geq\mu\left(\bigcup_{T\in\Gamma} T(F_1)\cap\mathring{F}_2\right)=\mu(\mathring{F}_2)=\mu(F_2).$$

Interchanging F_1 and F_2, we obtain $\mu(F_2) \geq \mu(F_1)$. Hence $\mu(F_2) = \mu(F_1)$. □

Thus we have proved a very important fact: the area of a fundamental region, if it is finite, is a numerical invariant of the group. An example of a Fuchsian group with a fundamental region of infinite area is the group generated by $z \to z+1$ (see also Example B above). Obviously, a compact fundamental region has finite area. Non-compact regions also may have finite area For example, for $\Gamma = PSL(2,\mathbb{Z})$, the fundamental region, which will be described in §3.2 (Example A), is a hyperbolic triangle with angles $\frac{\pi}{3}$, $\frac{\pi}{3}$, 0. By the Gauss-Bonnet formula (Theorem 1.4.2) its area is finite and is equal to $\pi - \frac{2\pi}{3} = \frac{\pi}{3}$.

THEOREM 3.1.2. *Let* Γ *be a discrete group of isometries of the upper half-plane* $\mathcal{H}$, *and* Λ *be a subgroup of* Γ *of index* n. *If*

$$\Gamma = \Lambda T_1 \cup \Lambda T_2 \cup \ldots \cup \Lambda T_n$$

is a decomposition of Γ *into* Λ*-cosets and if* F *is a fundamental region for* Γ *then*

(i) $F_1 = T_1(F) \cup T_2(F) \cup \ldots \cup T_n(F)$ *is a fundamental region for* Λ,

(ii) if $\mu(F)$ *is finite and the hyperbolic area of the boundary of* F *is zero then* $\mu(F_1) = n\mu(F)$.

PROOF of (i): Let $z \in \mathcal{H}$. Since F is a fundamental region for Γ, there exists $w \in F$ and $T \in \Gamma$ such that $z = T(w)$. We have $T = ST_i$ for some $S \in \Lambda$ and some i, $1 \leq i \leq n$. Therefore

$$z = ST_i(w) = S(T_i(w)).$$

Since $T_i(w) \in F_1$, z is in the Λ-orbit of some point of F_1. Hence the union of the Λ-images of F_1 is $\mathcal{H}$.

Now suppose that $z \in \mathring{F}_1$ and that $S(z) \in \mathring{F}_1$, for $S \in \Lambda$. We need to prove that $S = Id$. Let $\epsilon > 0$ be so small that $B_\epsilon(z)$ (the open hyperbolic disc of radius ϵ centered at z) is contained in $\mathring{F}_1$. Then $B_\epsilon(z)$ has a non-empty intersection with exactly k of the images of $\mathring{F}$ under $T_1, \ldots, T_n$, where $1 \leq k \leq n$. Suppose these images are $T_{i_1}(\mathring{F}), \ldots, T_{i_k}(\mathring{F})$. Let $B_\epsilon(S(z)) = S(B_\epsilon(z))$ have a non-empty intersection with $T_j(\mathring{F})$ say, $1 \leq j \leq n$. It follows that $B_\epsilon(z)$ has a non-empty intersection with $S^{-1}T_j(\mathring{F})$ so that $S^{-1}T_j = T_{i_l}$ where $1 \leq l \leq k$. Hence

$$\Lambda T_j = \Lambda S^{-1} T_j = \Lambda T_{i_l},$$

so that $T_j = T_{i_l}$ and $S = Id$. Hence $\mathring{F}_1$ contains precisely one point of each Λ-orbit.

PROOF of (ii): This follows immediately, as $\mu(T(F)) = \mu(F)$ for all $T \in PSL(2,\mathbb{R})$, and $\mu(T_i(F) \cap T_j(F)) = 0$ for $i \neq j$. □

3.2. The Dirichlet region

Let Γ be an arbitrary Fuchsian group and let $p \in \mathcal{H}$ be not fixed by any element of $\Gamma - \{Id\}$. Such points exist by Lemma 2.2.5. We define the *Dirichlet region for* Γ *centered at* p to be the set

$$D_p(\Gamma) = \{z \in \mathcal{H} \mid \rho(z,p) \leq \rho(z,T(p)) \text{ for all } T \in \Gamma\}. \qquad (3.2.1)$$

By the invariance of the hyperbolic metric under $PSL(2,\mathbb{R})$ this region can also be defined as

$$D_p(\Gamma) = \{z \in \mathcal{H} \mid \rho(z,p) \leq \rho(T(z),p) \text{ for all } T \in \Gamma\}. \qquad (3.2.2)$$

For each fixed $T_1 \in PSL(2,\mathbb{R})$,

$$\{z \in \mathcal{H} \mid \rho(z,p) \leq \rho(z,T_1(p))\} \qquad (3.2.3)$$

is the set of points z which are closer in the hyperbolic metric to p than to $T_1(p)$. Clearly, $p \in D_p(\Gamma)$ and as the Γ-orbit of p is discrete (Corollary 2.2.7), $D_p(\Gamma)$ contains a neighborhood of p. In order to describe the set (3.2.3) we join the points p and $T_1(p)$ by a geodesic segment and construct a line given by the equation

$$\rho(z,p) = \rho(z,T_1(p)).$$

DEFINITION. A *perpendicular bisector* of the geodesic segment $[z_1,z_2]$ is the unique geodesic through w, the mid-point of $[z_1,z_2]$ orthogonal to $[z_1,z_2]$ (Fig. 8).

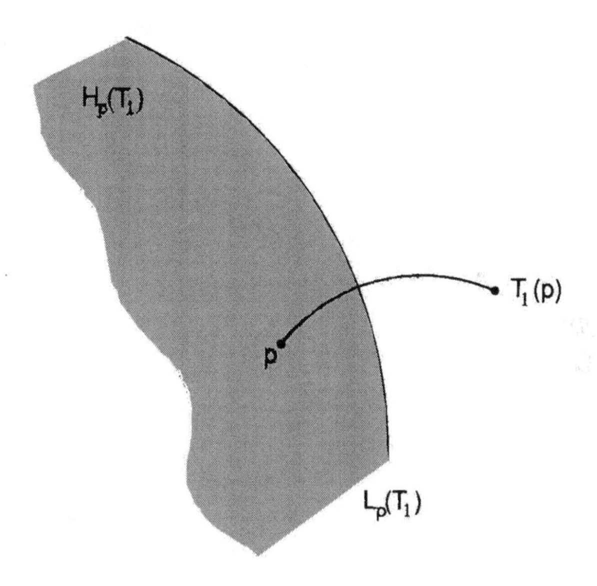

Fig. 8

LEMMA 3.2.1. *A line given by the equation*

$$\rho(z,z_1) = \rho(z,z_2) \tag{3.2.4}$$

is the perpendicular bisector of the geodesic segment $[z_1, z_2]$.

PROOF: We may assume that $z_1=i$, $z_2=ir^2$ with $r>0$: thus $w=ir$ and the perpendicular bisector is given by the equation $|z|=r$. On the other hand, by Theorem 1.2.6(ii) (3.2.4) is equivalent to

$$\frac{|z-z_1|^2}{y}=\frac{|z-z_2|^2}{r^2y}$$

which simplifies to $|z|=r$. □

We shall denote the perpendicular bisector of the geodesic segment $[p,T_1(p)]$ by $L_p(T_1)$, and the hyperbolic half-plane containing p described in (3.2.3) by $H_p(T_1)$ (see Fig. 8). Thus $D_p(\Gamma)$ is the intersection of hyperbolic half-planes:

$$D_p(\Gamma)=\bigcap_{T\in\Gamma,\ T\neq Id} H_p(T),$$

and thus is a *hyperbolically convex region.*

THEOREM 3.2.2. *If* p *is not fixed by any element of* $\Gamma-\{Id\}$, *then* $D_p(\Gamma)$ *is a connected fundamental region for* Γ.

PROOF: Let $z\in\mathcal{H}$, and Γz be its Γ-orbit. Since Γz is a discrete set, there exists $z_0\in\Gamma z$ with the smallest $\rho(z_0,p)$. Then $\rho(z_0,p)\leq\rho(T(z_0),p)$ for all $T\in\Gamma$, and by (3.2.2) $z_0\in D_p(\Gamma)$. Thus $D_p(\Gamma)$ contains at least one point from every Γ-orbit.

Next we show that if z_1, z_2 are in the interior of $D_p(\Gamma)$, they cannot lie in the same Γ-orbit. If $\rho(z,p)=\rho(T(z),p)$ for some $T\in\Gamma-\{Id\}$, then $\rho(z,p)=\rho(z,T^{-1}(p))$ and hence $z\in L_p(T^{-1})$. Then either $z\notin D_p(\Gamma)$ or z lies on the boundary of $D_p(\Gamma)$; hence if z is in the interior of $D_p(\Gamma)$, $\rho(z,p)<\rho(T(z),p)$ for all $T\in\Gamma-\{Id\}$. If two points z_1, z_2 lie in the same Γ-orbit, this implies $\rho(z_1,p)<\rho(z_2,p)$ and $\rho(z_2,p)<\rho(z_1,p)$, a contradiction. Thus the interior of $D_p(\Gamma)$ contains at most one point in each Γ-orbit. Being an intersection of closed half-planes, $D_p(\Gamma)$ is closed and convex. Thus $D_p(\Gamma)$ is path-connected, hence connected. □

EXAMPLE A. $\Gamma=PSL(2,\mathbb{Z})$. It is easily verified that ki (k>1) is not fixed by any non-identity element of the modular group, so choose p=ki, where k>1. We shall show that the region

$$F=\{z\in\mathcal{H} \mid |z|\geq 1,\ |Re(z)|\leq\tfrac{1}{2}\},$$

illustrated in Figure 9 is the Dirichlet region for Γ centered at p.

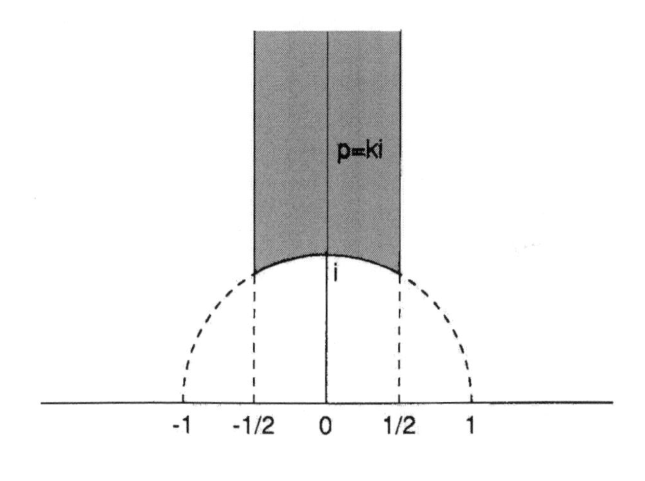

Fig. 9

First, the isometries $T(z)=z+1$, $S(z)=-1/z$ are in Γ; and, as can be easily verified, the three geodesic sides of F are $L_p(T)$, $L_p(T^{-1})$ and $L_p(S)$. This shows that $D_p(\Gamma)\subset F$. If $D_p(\Gamma)\neq F$, there exists $z\in\mathring{F}$ and $h\in\Gamma$ such that $h(z)\in\mathring{F}$. We shall now show that this cannot happen. Suppose that

$$h(z)=\frac{az+b}{cz+d},\quad (a,\ b,\ c,\ d\in\mathbb{Z},\ ad-bc=1).$$

Then

$$|cz+d|^2=c^2|z|^2+2\mathrm{Re}(z)cd+d^2>c^2+d^2-|cd|=(|c|-|d|)^2+|cd|,$$

since $|z|>1$ and $\mathrm{Re}(z)>-\frac{1}{2}$. This lower bound is an integer: it is non-negative and is not zero (this would be possible only if $c=d=0$, which contradicts $ad-bc=1$). Therefore it is at least 1 and $|cz+d|>1$. Hence

$$\mathrm{Im}\ h(z)=\frac{\mathrm{Im}(z)}{|cz+d|^2}<\mathrm{Im}(z).$$

Exactly the same argument holds with z, h replaced by $h(z)$, h^{-1}, and a contradiction is reached: thus $D_p(\Gamma)=F$. □

3.3. Isometric circles and the Ford fundamental region

Let

$$T(z)=\frac{az+b}{cz+d}\in PSL(2,\mathbb{R}). \tag{3.3.1}$$

Since $T'(z)=(cz+d)^{-2}$, the Euclidean lengths are multiplied by $|T'(z)|=|cz+d|^{-2}$. An infinitesimal region is carried into a similar region with lengths multiplied by $|cz+d|^{-2}$. The Euclidean area therefore is multiplied by $|cz+d|^{-4}$. The Euclidean lengths and areas are unaltered in magnitude if and only if $|cz+d|=1$. If $c\neq 0$, the locus of such z is a circle

$$\left|z+\frac{d}{c}\right|=\frac{1}{|c|}$$

with center at $-\frac{d}{c}$ and radius $\frac{1}{|c|}$.

DEFINITION. Let T be as in (3.3.1) with $c \neq 0$. The circle

$$I(T)=\{z \in \mathbb{C} \mid |cz+d|=1\},$$

which is the complete locus of points where the transformation T acts as a Euclidean isometry, is called the *isometric circle* of the transformation T.

REMARKS: 1. Note that if $c=0$, there is no unique circle with the isometric property. In this case, ∞ is a fixed point. If a transformation T is parabolic, e.g. $z \to z+1$, all Euclidean lengths are unaltered. If a transformation is hyperbolic, e.g. $z \to \lambda z$ $(\lambda \neq 1)$, all Euclidean lengths are altered.

2. The definition is valid for the unit disc model. In this case, instead of $PSL(2, \mathbb{R})$, we have (see Exercise 1.10) the group of orientation-preserving isometries of the unit disc $\mathcal{U}$ given by the matrices

$$T(z)=\frac{az+\bar{c}}{cz+\bar{a}} \quad (a,c \in \mathbb{C},\ a\bar{a}-c\bar{c}=1). \tag{3.3.2}$$

If $c=0$, then $T(z)=e^{i\phi}z$, an elliptic transformation fixing the origin. All Euclidean lengths in this case are unaltered.

We shall denote the set of points inside of the isometric circle $I(T)$ by $\check{I}(T)$, and the set of points outside of $I(T)$ by $\hat{I}(T)$.

THEOREM 3.3.1. *The transformation* T *increases Euclidean lengths and areas inside of the isometric circle* I(T), *and decreases them outside of the isometric circle* I(T).

PROOF: Let $z \in \check{I}(T)$. Hence $\left|z+\frac{d}{c}\right|<\frac{1}{|c|}$, i.e. $|cz+d|<1$, which implies $|T'(z)|>1$. Similarly, $z \in \hat{I}(T)$ implies $|T'(z)|<1$. □

THEOREM 3.3.2. *The isometric circles* I(T) *and* $I(T^{-1})$ *have the same radius; and* I(T) *is carried into* $I(T^{-1})$ *by the transformation* T.

PROOF: We have $T^{-1}(z)=\frac{-dz+b}{cz-a}$. The isometric circle $I(T^{-1})$ is given by the equation $|cz-a|=1$, and has center at $\frac{a}{c}$ and radius $\frac{1}{|c|}$, equal to the radius of I(T). T carries I(T) into a circle I_0 outside of alteration of Euclidean lengths. Hence T^{-1} carries I_0 back to I(T) without alteration of Euclidean lengths. But $I(T^{-1})$ is the complete locus of points in the neighborhood of which T^{-1} effects no change of Euclidean lengths. Therefore $I^0=I(T^{-1})$. □

THEOREM 3.3.3. *Isometric circles are geodesics in* $\mathcal{H}$.

PROOF: Let T be of the form (3.3.1). Then the center of I(T) is $-\frac{d}{c} \in \mathbb{R}$. Therefore I(T) is orthogonal to the real axis. □

REMARK: This property is preserved as we switch to the unit disc model, although the centers of the isometric circles no longer belong to the principal circle Σ (see Exercise 3.3).

There is an intimate relationship between fractional linear transformations and the geometrical transformation called *inversion in a circle.*

DEFINITION. Let Q be a circle in $\mathbb{R}^2$ with center K and radius r.

Given any point $P \neq K$ in $\mathbb{R}^2$, a point P_1 is called *inverse* to P if

$$\begin{aligned}&\text{(i)}\quad P_1 \text{ lies on the ray from K to P,}\\&\text{(ii)}\quad KP_1 \cdot KP = r^2.\end{aligned} \tag{3.3.3}$$

The relationship is a reciprocal one: if P_1 is inverse to P, P is inverse to P_1. We say that P and P_1 are *inverse with respect to* Q.

Let P, P_1 and K be the points z, z_1 and k in $\mathbb{C}$. (3.3.3) can be rewritten as

$$|(z_1-k)(z-k)|=r^2, \ \arg(z_1-k)=\arg(z-k).$$

Since $\arg(z-k)=-\arg(\bar{z}-\bar{k})$, the two equations are satisfied if and only if

$$(z_1-k)(\bar{z}-\bar{k})=r^2. \tag{3.3.4}$$

So we get the formula for inversion in a circle:

$$z_1=\frac{k\bar{z}+r^2-|k|^2}{\bar{z}-\bar{k}}. \tag{3.3.5}$$

If a circle is centered at k=0 and has radius 1, the formula looks especially simple:

$$z_1=\frac{1}{\bar{z}}. \tag{3.3.6}$$

Let us now look at the unit disc model, $\mathcal{U}$.

THEOREM 3.3.4. *Any orientation-preserving isometry* T *of the unit disc* $\mathcal{U}$ *is an inversion in* I(T) *followed by a reflection in the straight line* L, *the Euclidean bisector between the centers of the isometric circles* I(T) *and* $I(T^{-1})$.

PROOF: We have $T(z)=\frac{az+\bar{c}}{cz+\bar{a}}$, $a,c \in \mathbb{C}$, $a\bar{a}-c\bar{c}=1$. Since the isometric circles $I(T)$ and $I(T^{-1})$ have the same radius (Theorem 3.3.2) and they both are orthogonal to the principal circle Σ (Theorem 3.3.3), they are symmetric with respect to a straight line L passing through the center of Σ. Suppose $c=re^{i\phi}$, then let $T_\phi z=e^{i\phi}z$, and $S=T_\phi \circ T \circ T_\phi^{-1}$. Then the center of $I(S)$ is the point $\frac{a}{r}$, and the center of $I(S^{-1})$ is the point $-\frac{\bar{a}}{r}$ which are symmetric with respect to the imaginary axis. Suppose we proved that $S=s\circ\mathfrak{J}$ where $\mathfrak{J}$ is inversion in $I(S)$ and s is the reflection in the imaginary axis. Then

$$T=T_\phi^{-1}\circ s\circ T_\phi\circ T_\phi^{-1}\circ\mathfrak{J}\circ T_\phi,$$

where $T_\phi^{-1}\circ s\circ T_\phi$ is the symmetry in L, and $T_\phi^{-1}\circ\mathfrak{J}\circ T_\phi$ is the inversion in $I(T)$. It is sufficient therefore to prove the theorem assuming that the centers of $I(T)$ and $I(T^{-1})$ ($-\frac{\bar{a}}{\bar{c}}$ and $\frac{a}{c}$, respectively) are symmetric with respect to the imaginary axis. In this case, $-\frac{\bar{a}}{\bar{c}}=-\frac{\bar{a}}{c}$, which implies $c=\bar{c}$. Using (3.3.5) for the inversion in $I(T)$, we obtain

$$z_1=-\frac{\frac{\bar{a}}{c}\bar{z}+1}{\bar{z}+\frac{a}{c}}.$$

Applying a reflection with respect to the imaginary axis, we obtain

$$z'=-\bar{z}_1=\frac{\frac{a}{c}z+1}{z+\frac{\bar{a}}{c}}=\frac{az+\bar{c}}{cz+\bar{a}}=T(z). \qquad \square$$

In the rest of this section, Γ will be a discrete group of orientation-preserving isometries of the unit disc $\mathcal{U}$ (sometimes also referred to as a Fuchsian group). We assume that 0 is not an elliptic fixed point, i.e. that $c\neq 0$ for all $T(z)=\frac{az+\bar{c}}{cz+\bar{a}}$ in the group Γ. We define

$$R_0=\overline{\bigcap_{T\in\Gamma}\hat{I}(T)\bigcap\mathcal{U}},$$

the closure of the set of points in $\mathcal{U}$ which are exterior to the isometric circles of all transformations in the group Γ. We shall prove that R_0 is a fundamental region for Γ, called the *Ford fundamental region.*

THEOREM 3.3.5. R_0 *is a fundamental region for* Γ.

PROOF: We shall prove that R_0 is a Dirichlet region $D_0(\Gamma)$, and the theorem will follow from Theorem 3.2.2.

LEMMA 3.3.6. *Let* Σ *be the principal circle centered at* q, $\Sigma=\partial\mathfrak{U}$, $T\in\Gamma$, $h=d(z,q)$, *and* $h'=d(T(z),q)$ *(where* d *denotes the Euclidean distance in* $\mathfrak{U}$*). Then:*
i) $h'=h$ *if* $z\in I(T)$ *or* $z\in\Sigma$.
ii) $h'<h$ *if* $z\in\check{I}(T)\cap\mathfrak{U}$,
iii) $h'>h$ *if* $z\in\hat{I}(T)\cap\mathfrak{U}$.

PROOF of Lemma: According to Theorem 3.3.4, T is an inversion in I(T) followed by a reflection in a certain line L passing through q. Obviously, the reflection does not alter distances from q. The lemma hinges, then, on what happens when z is inverted in I(T).

The relative magnitudes of the distances of a point and its inverse are independent of the scale used and the rigid motions of the plane. If we change the scale via a transformation $z\rightarrow rz$ $(r>0)$, both Σ and I(T) will be transformed into circles orthogonal to each other, and we can choose r in such a way that the radius of I(T) becomes equal to 1 (the radius of Σ becomes equal to r). Via a rigid motion of the plane we achieve the situation in which I(T) is the unit circle $z\bar{z}=1$ and q lies on the real axis. The equation of Σ is $(z-q)(\bar{z}-q)=r^2$, where for orthogonality, $r^2+1=q^2$; whence,

$$z\bar{z}-q(z+\bar{z})+1=0.$$

The expression on the left-hand side of this equation is negative for points inside $\mathfrak{U}$. Now

$$h^2=(z-q)(\bar{z}-q)=z\bar{z}-q(z+\bar{z})+1+r^2,$$

and since $z_1=1/\bar{z}$ (see (3.3.5)),

$$h'^2=z_1\bar{z}_1-q(z_1+\bar{z}_1)+1+r^2=\frac{1-q(\bar{z}+z)+z\bar{z}}{z\bar{z}}+r^2;$$

whence,

$$h^2-h'^2=\frac{(z\bar{z}-1)(z\bar{z}-q(z+\bar{z})+1)}{z\bar{z}}. \tag{3.3.7}$$

The lemma follows immediately from equation (3.3.7). If z is inside I(T), the factors in the numerator of the right-hand side of (3.3.7) are both negative, hence $h'<h$. If z is outside I(T), the factors differ in sign, hence $h'>h$. If $z\in\Sigma$ or $z\in I(T)$, $h'=h$. □

To finish the proof of Theorem 3.3.5, we apply Lemma 3.3.6 with $q=0$ and $z\in R_0$. We have

$$d(T(z),0)\geq d(z,0)$$

for all $T\in\Gamma$. By Exercise 1.6, $\rho(0,z)=\ln\frac{1+d(z,0)}{1-d(z,0)}$, hence it is a monotone increasing function of $d(z,0)$. Therefore, for $z\in R_0$ we have

$$\rho(z,0)\leq\rho(T(z),0)$$

for all $T\in\Gamma$; then by (3.2.2), $R_0=D_0(\Gamma)$. □

THEOREM 3.3.7. *Given any infinite sequence of distinct isometric circles* I_1, I_2, ... *of transformations of the group* Γ *with radii* $r_1, r_2, \ldots$, *we have* $\lim_{n\to\infty} r_n=0$.

PROOF: The transformations are of the form

$$T(z)=\frac{az+\bar{c}}{cz+\bar{a}} \quad (a,c\in\mathbb{C},\ |a|^2-|c|^2=1). \tag{3.3.8}$$

Recall that the radius of I(T) is equal to $\frac{1}{|c|}$. Let $\epsilon>0$ be given. There are only finitely many $T \in \Gamma$ with $|c|<1/\epsilon$. This follows from the discreteness of Γ and the relation $|a|^2-|c|^2=1$. Hence there are only finitely many $T \in \Gamma$ with $I(T)$ of radius exceeding ϵ, and the theorem follows. □

3.4. The limit set of Γ

We saw at the end of §2.2 that for any Fuchsian group Γ, the limit set $\Lambda(\Gamma) \subseteq \mathbb{R} \cup \{\infty\}$; or, in the unit disc model, $\Lambda(\Gamma)$ is a subset of the principal circle Σ. A point which is not a limit point is called an *ordinary point.*

Consider the set of the centers of the isometric circles of all elements in Γ, and denote the set of its limit points by $\Lambda_0(\Gamma)$. If the group Γ contains an infinite number of elements, $\Lambda_0(\Gamma) \neq \emptyset$. In the upper half-plane model, the centers of all isometric circles belong to the real axis, hence $\Lambda_0(\Gamma) \subseteq \mathbb{R} \cup \{\infty\}$.

THEOREM 3.4.1. *In the unit disc model,* $\Lambda_0(\Gamma)$ *is a subset of the principal circle* Σ.

PROOF: Since the isometric circles are orthogonal to Σ, their centers lie outside of Σ. Suppose there is a point $\delta \in \Lambda_0(\Gamma)$ not belonging to Σ, then there exists a circle centered at δ which lies outside of Σ. Take a sequence of isometric circles I_n with radii r_n and centers q_n such that $q_n \to \delta$. Using Theorem 3.3.7, we conclude that I_n will not intersect Σ for large enough n, which leads to a contradiction. □

THEOREM 3.4.2. $\Lambda(\Gamma)=\Lambda_0(\Gamma)$.

PROOF: We first prove that $\Lambda_0(\Gamma) \subseteq \Lambda(\Gamma)$. Let $\alpha \in \Lambda_0(\Gamma)$. Then there exists a sequence of distinct transformations $\{S_n\} \in \Gamma$ such that the centers of $I(S_n^{-1})$, $p_n \to \alpha$. Let $\epsilon > 0$ be given. Using Theorem 3.3.7, we conclude that there exists $N > 0$ such that for $n > N$ and $z \in \check{I}(S_n^{-1})$, $d(\alpha, z) < \epsilon$. Let B be a closed disc of radius $1 - \epsilon$ concentric with the principle circle. Let $\{q_n\}$ be the sequence of centers of $I(S_n)$. According to Theorem 3.3.7, it is bounded in $\mathbf{C}$, and hence contains a converging subsequence, so we may assume $q_n \to \beta$, taking a subsequence if necessary. Since $\beta \in \Sigma$ (Theorem 3.4.1) and using Theorem 3.3.7 again, we conclude that there exists $M > 0$ such that for $n > M$ and $z \in \check{I}(S_n)$, $d(\beta, z) < \epsilon$, i.e.

$$\bigcup_{n>M} I(S_n) \cap B = \emptyset.$$

Let $z \in B$. Then z lies outside of $I(S_n)$ for all $n > M$, hence $S_n(z) \in \check{I}(S_n^{-1})$. So we obtain for $n > \max(N, M)$, $d(S_n(z), \alpha) < \epsilon$, and conclude that for any $z \in B$, $S_n(z) \to \alpha$.

Now we prove that $\Lambda(\Gamma) \subseteq \Lambda_0(\Gamma)$. Let $\alpha \in \Lambda(\Gamma)$, then there exists $z \in \mathfrak{U}$ and a sequence $\{T_n\}$ of distinct transformations in Γ such that $T_n z \to \alpha$. The sequence of centers of $I(T_n)$, $\{q_n\}$, has a subsequence converging to a point $\beta \in \Lambda_0(\Gamma) \subseteq \Sigma$. Thus we may assume, taking a subsequence if necessary, that z lies outside of all $I(T_n)$, hence $T_n(z) \in \check{I}(T_n^{-1})$. Let the radii of the isometric circles of the transformations T_n^{-1} be equal to r_n. By Theorem 3.3.7, $r_n \to 0$. Let the centers of $I(T_n^{-1})$ be p_n, and $\epsilon > 0$ be given. There exists $N > 0$ such that for $n > N$, $r_n < \epsilon/2$ and $d(T_n(z), \alpha) < \epsilon/2$. Then for $n > N$,

$$d(p_n, \alpha) \leq d(p_n, T_n(z)) + d(T_n(z), \alpha) \leq \epsilon.$$

Hence $p_n \to \alpha$, and $\alpha \in \Lambda_0(\Gamma)$. □

LEMMA 3.4.3. *Let α, β, $\delta \in \Sigma$ be three distinct points with $\alpha \in \Lambda(\Gamma)$. Then α is a limit point either for $\Gamma\beta$ or for $\Gamma\delta$.*

PROOF: We may assume, taking subsequences if necessary, that for a sequence of distinct transformations in Γ, $\{T_n\}$, the centers of the isometric circles $I(T_n^{-1})$, $p_n \to \alpha \in \Lambda(\Gamma)$. Suppose the centers of the isometric circles $I(T_n)$, $q_n \to \gamma \in \Lambda(\Gamma)$. Notice that γ may coincide with β or δ but not with both. Suppose $\delta \neq \gamma$. Since the radii of isometric circles tend to 0 (Theorem 3.3.7), δ lies outside of infinitely many isometric circles $I(T_n)$, hence $T_n(\delta) \in \check{I}(T_n^{-1})$. Therefore $T_n(\delta) \to \alpha$, i.e. α is a limit point for $\Gamma\delta$. □

REMARK: Lemma 3.4.3 is valid also for the upper half-plane model since $f^{-1}(\Lambda(\Gamma)) = \Lambda(f^{-1} \circ \Gamma \circ f)$, and α is a limit point for $\Gamma\beta$ if and only if $f^{-1}(\alpha)$ is a limit point for $f^{-1} \circ \Gamma \circ f(f^{-1}\beta)$, where f is as in (1.2.2).

THEOREM 3.4.4. *If $\Lambda(\Gamma)$ contains more than one point, it is the closure of the set of fixed points of the hyperbolic transformations of Γ.*

PROOF: This theorem is true for both models. First we show that Γ must contain at least one hyperbolic element. We may use the upper half-plane model since the trace is invariant under conjugations, hence the type of an element does not depend on the choice of the model. Assume Γ contains only elliptic and parabolic elements. Γ does not consist of only elliptic elements, for then it would be a finite cyclic group (Corollary 2.4.2) and $\Lambda(\Gamma) = \emptyset$. Let T be a parabolic element whose fixed point may be assumed to be ∞. Then T is a translation $z \to z+k$. The group Γ must contain an element $S = \dfrac{az+b}{cz+d}$ which does not fix ∞; otherwise Γ would be an infinite cyclic parabolic group with $\Lambda(\Gamma) = \{\infty\}$. Hence $c \neq 0$ and $|\mathrm{tr}\, T^n S| = |a+d+nkc| > 2$ if n is sufficiently large. It follows that $T^n S$ is hyperbolic, a contradiction.

In order to see that $\Lambda(\Gamma)$ is closed, we use the unit disc model. According to the preceding remark, it will follow that $\Lambda(f^{-1}\circ\Gamma\circ f)$ is also closed as the image of a closed set under a continuous map. Since each point of $\Lambda(\Gamma)$ has an infinite number of centers of isometric circles in its neighborhood, a limit point of $\Lambda(\Gamma)$ has also an infinite number of centers of isometric circles in its neighborhood. Thus it belongs to $\Lambda(\Gamma)$, so $\Lambda(\Gamma)$ is closed. $\Lambda(\Gamma)$ contains hyperbolic fixed points for all hyperbolic elements in Γ, and therefore it contains the closure of the set of hyperbolic fixed points. Suppose now that $\alpha\in\Lambda(\Gamma)$ and α is not a hyperbolic fixed point; we must show that α is a limit point of hyperbolic fixed points. We have seen that Γ has at least one hyperbolic element. Let μ_1 and μ_2 be two hyperbolic fixed points. By Lemma 3.4.3, α is a limit point for either $\Gamma\mu_1$ or for $\Gamma\mu_2$. Since the image of a fixed point is itself a fixed point, the theorem is established. □

THEOREM 3.4.5. *The limit set* $\Lambda(\Gamma)$ *is* Γ*-invariant.*

PROOF: Let $\alpha\in\Lambda(\Gamma)$. Then there is $z\in\mathfrak{U}$ and a sequence of distinct transformations in Γ, $\{T_n\}$, such that $T_n z\to\alpha$. Let $S\in\Gamma$. Then $(ST_nS^{-1})(S(z))\to S(\alpha)$, and hence $S(\alpha)\in\Lambda(\Gamma)$. □

THEOREM 3.4.6. *If the set* $\Lambda(\Gamma)$ *contains more than two points, then either*

(i) $\Lambda(\Gamma)=\Sigma$, *or*

(ii) $\Lambda(\Gamma)$ *is a perfect nowhere dense subset of* Σ.

PROOF: We have seen already in the proof of Theorem 3.4.4 that $\Lambda(\Gamma)$ is closed. In order to see that it is a perfect set, we have to show that each point of $\Lambda(\Gamma)$ is a limit point of points of the set $\Lambda(\Gamma)$. Let α, β, $\delta\in\Lambda(\Gamma)$ be three distinct limit points. By Lemma 3.4.3, α is a limit point for either $\Gamma\beta$ or $\Gamma\delta$. Such points belong to $\Lambda(\Gamma)$ by Theorem 3.4.5, hence $\Lambda(\Gamma)$ is perfect.

We must show now that unless every point of Σ belongs to $\Lambda(\Gamma)$, $\Lambda(\Gamma)$ is nowhere dense. Let $z_0 \in \Sigma$, $z_0 \notin \Lambda(\Gamma)$. Since $\Lambda(\Gamma)$ is closed, its complement is open, and there exists a neighborhood of z_0 consisting only of ordinary points. Therefore, there is an arc in Σ, h, containing z_0 and consisting of ordinary points. Let $\alpha \in \Lambda(\Gamma)$. By Lemma 3.4.3, there exists $z_1 \in h$ and a sequence $\{T_n\}$ such that $T_n z_1 \to \alpha$. Since an image of an ordinary point is an ordinary point (Theorem 3.4.5), we conclude that there are ordinary points in each neighborhood of Σ, hence $\Lambda(\Gamma)$ is nowhere dense in Σ. □

According to Theorem 3.4.6, we shall classify Fuchsian groups as follows:

(a) *Fuchsian groups of the first kind*, or groups for which every point of the principal circle is a limit point.

(b) *Fuchsian groups of the second kind*, or groups whose limit points are nowhere dense on the principal circle.

REMARK: According to Theorem 3.4.6, the limit set of a Fuchsian group of the second kind can be one of the following: an empty set, a set containing one or two points, or a perfect (and therefore infinite) nowhere dense set.

3.5. Structure of a Dirichlet region

Dirichlet regions for Fuchsian groups can be quite complicated. They are bounded by geodesics in $\mathcal{H}$ and possibly by segments of the real axis. If two such geodesics intersect in $\mathcal{H}$, their point of intersection is called a *vertex* of the Dirichlet region. It can be shown that the vertices are isolated (see Exercise 3.10 below) so that a Dirichlet region is bounded by a union of (possibly infinitely many) geodesics and possibly segments of the real axis (see Fig. 10 for the unit disc model).

We shall be interested in the tessellation of $\mathcal{H}$ formed by a Dirichlet region F and all its images under Γ (called *faces*): $\{T(F) \mid T \in \Gamma\}$. This tesselation will be referred to as a *Dirichlet tessellation.* (See Fig. 11 for a Dirichlet tessellation for the modular group.) The next theorem shows that the Dirichlet tessellation has nice local properties. Recall the definition of locally finite family of subsets from § 2.2.

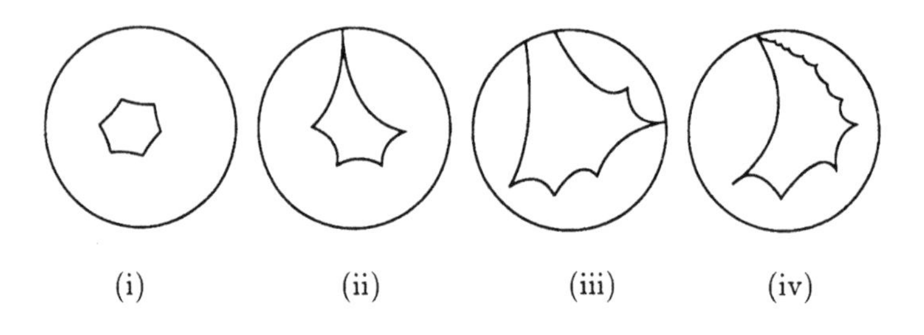

Fig. 10

DEFINITION. A fundamental region F for a Fuchsian group Γ is called *locally finite* if the tessellation $\{T(F) \mid T \in \Gamma\}$ is locally finite.

THEOREM 3.5.1. *A Dirichlet region is locally finite.*

PROOF: Let $F=D_p(\Gamma)$ where p is not fixed by any element of $\Gamma-\{Id\}$. Let $a\in F$, and let $K\subset\mathcal{H}$ be a compact neighborhood of a.

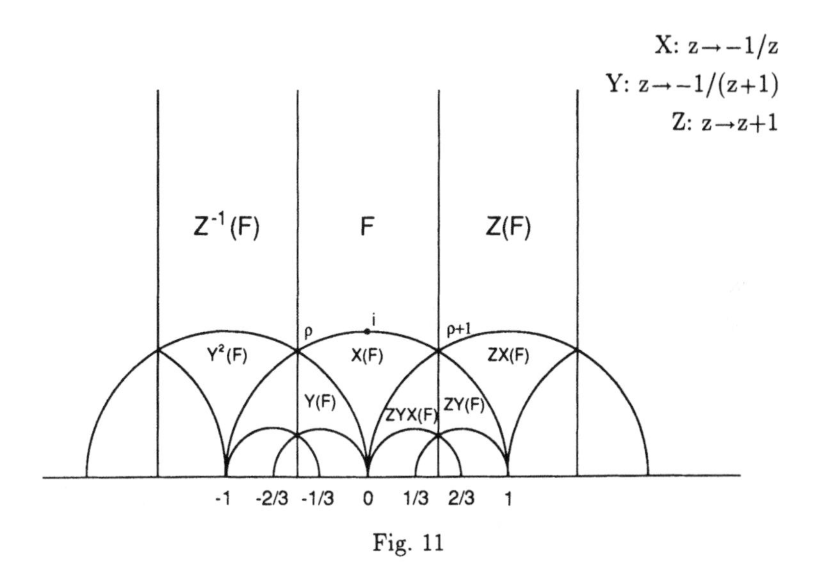

Fig. 11

Suppose that $K\cap T_i(F)\neq\emptyset$ for some infinite sequence T_1, T_2, ... of distinct elements of Γ. Let $\sigma=\sup_{z\in K}\rho(p,z)$. Since $\sigma\leq\rho(p,a)+\rho(a,z)$, for all $z\in K$, and K is bounded, σ is finite. Let $w_j\in K\cap T_j(F)$. Then $w_j=T_j(z_j)$ for $z_j\in F$, and by the triangle inequality,

$$\begin{aligned}\rho(p,T_j(p)) &\leq\rho(p,w_j)+\rho(w_j,T_j(p))\\ &=\rho(p,w_j)+\rho(z_j,p)\\ &\leq\rho(p,w_j)+\rho(w_j,p)\quad(\text{as } z_j\in D_p(\Gamma))\\ &\leq 2\sigma.\end{aligned}$$

Thus the infinite set of points $T_1(p)$, $T_2(p)$, ... belongs to the compact hyperbolic ball with center p and radius 2σ, but this contradicts the properly discontinuous action of Γ. □

We call two points $u,v \in \mathcal{H}$ *congruent* if they belong to the same Γ-orbit. First, notice that two points in a fundamental region F may be congruent only if they belong to the boundary of F. Suppose now that F is a Dirichlet region for Γ, and let us consider congruent vertices of F. The congruence is an equivalence relation on the vertices of F and the equivalence classes are called *cycles.* If u is fixed by an elliptic element S, then $v=Tu$ is fixed by the elliptic element TST^{-1}. Thus if one vertex of the cycle is fixed by an elliptic element, then all the vertices of that cycle are fixed by conjugate elliptic elements. Such a cycle is called an *elliptic cycle* and the vertices are called *elliptic vertices.* The number of elliptic cycles is equal to the number of non-congruent elliptic points in F.

Since the Dirichlet region F is a fundamental region, it is clear that every point $w \in \mathcal{H}$ fixed by an elliptic element S' of Γ lies on the boundary of T(F) for some $T \in \Gamma$. Hence $u=T^{-1}(w)$ lies on the boundary of F and is fixed by the elliptic element $S=T^{-1}S'T$. By Theorem 2.3.7, S has finite order k. Suppose first that $k \geq 3$: then as S is an isometry fixing u which maps geodesics to geodesics, u must be a vertex whose angle θ is at most $2\pi/k$. (See Fig. 11 where the angle at the elliptic fixed point ρ of order 3 is $2\pi/6$.) The hyperbolically convex region F is bounded by a union of geodesics. The intersection of F with these geodesics is either a single point or a segment of a geodesic. These segments are called *sides* of F. If S has order 2, its fixed point might lie on the interior of a side of F. In this case, S interchanges the two segments of this side separated by the fixed point. We will include such elliptic fixed points as vertices of F, the angle at such vertex being π. Thus a *vertex* of F is a point of intersection in $\mathcal{H}$ of two bounding geodesics of F or a fixed point of an elliptic element of order 2. (All the previous definitions such as conjugate, elliptic cycles, etc. apply to this extended set of vertices.)

If a point in $\mathcal{H}$ has a nontrivial stabilizer in Γ, this stabilizer is a finite cyclic subgroup of Γ by Theorem 2.2.6; it is a *maximal finite cyclic subgroup* of Γ by Lemma 2.3.1. Conversely, every maximal finite cyclic subgroup of Γ is a stabilizer of a single point in $\mathcal{H}$. We can summarize the above as:

THEOREM 3.5.2. *There is a one-to-one correspondence between the elliptic cycles of* F *and the conjugacy classes of non-trivial maximal finite cyclic subgroups of* Γ. □

EXAMPLE A. Let Γ be the modular group. The Dirichlet region F in Figure 9 has vertices in $\mathcal{H}$ at $\rho=\frac{-1+i\sqrt{3}}{2}$, $\rho+1=\frac{1+i\sqrt{3}}{2}$ and i. These are stabilized by the cyclic subgroups generated by $z\to\frac{-z-1}{z}$, $z\to\frac{z-1}{z}$, and $z\to-\frac{1}{z}$, respectively. The verteces ρ and $\rho+1$ belong to the same cycle since they are congruent via $z\to z+1$. Each of them is fixed by an elliptic element of order 3. By Exercise 3.12, these two verteces form an elliptic cycle. The point i is fixed by an elliptic element of order 2. It follows from Exercise 3.12 that i is the only such point. Thus $\{i\}$ is an elliptic cycle consisting of just one vertex. By Theorem 3.5.2, the modular group has two conjugacy classes of maximal finite cyclic subgroups, one consisting of groups of order 2, the other consisting of groups of order 3.

DEFINITION. The orders of non-conjugate maximal finite cyclic subgroups of Γ are called the *periods* of Γ.

Each period is repeated as many times as there are conjugacy classes of maximal finite cyclic subgroups of that order. Thus the modular group has periods 2, 3.

A parabolic element can be considered as an elliptic element of infinite order; it has a unique fixed point in $\mathbb{R}\bigcup\{\infty\}$. Hence if a point in $\mathbb{R}\bigcup\{\infty\}$ has a non-trivial stabilizer in Γ, it is a *maximal (cyclic) parabolic subgroup* of Γ, and every maximal parabolic subgroup of Γ is a stabilizer of a single point in $\mathbb{R}\bigcup\{\infty\}$. Let F be a Dirichlet region for Γ with parabolic elements. It will be shown in § 4.2 that in this case F is not compact (Theorem 4.2.1), and if additionally $\mu(F)<\infty$, then F has at least one *vertex at infinity*, i.e. two bounding geodesics of F meet there (Theorem 4.2.2). Moreover, each vertex at infinity is a parabolic fixed point for a maximal parabolic subgroup of Γ (Theorem 4.2.5), and non-congruent verteces at infinity of F are in a one-to-one correspondence with conjugasy classes of maximal parabolic subgroups of Γ (Corollary 4.2.6). If we allow infinite periods, the period ∞ will occur the same number of times as there are conjugacy classes of maximal parabolic subgroups. This number is called the *parabolic class number of* Γ. It is easily calculated that in the modular group every parabolic element is conjugate to $z\to z+n$ for some $n\in\mathbb{Z}$, so that the modular group has periods 2, 3, ∞. The angle at a vertex at infinity is 0. With this convention, the Dirichlet region for the modular group described in §3.2 has a vertex at ∞ whose angle is $\frac{\pi}{\infty}=0$.

The following result relates the sum of angles at all elliptic verticies belonging to an elliptic cycle with the order of that cycle.

THEOREM 3.5.3. *Let* F *be a Dirichlet region for* Γ. *Let* $\theta_1, \theta_2, \dots, \theta_t$ *be the internal angles at all congruent vertices of* F. *Let* m *be the order of the stabilizer in* Γ *of one of these vertices. Then* $\theta_1+\dots+\theta_t=2\pi/m$.

REMARKS: 1. As F is locally finite, there are only finitely many vertices in a congruent cycle.

2. As the stabilizers of two points in a congruent set are conjugate

subgroups of Γ, they have the same order.

3. If a vertex is not a fixed point, we have $m=1$ and $\theta_1+...+\theta_t=2\pi$.

PROOF: Let $v_1, \ldots, v_t$ be the vertices of the congruent set, the internal angles being $\theta_1, \ldots, \theta_t$. Let

$$H=\{Id, S, S^2, \ldots, S^{m-1}\}$$

be the stabilizer of v_1 in Γ. Then each $S^r(F)$ $(0\le r\le m-1)$ has a vertex at v_1 whose angle is θ_1. Suppose $T_k(v_k)=v_1$ for some $T_k\in\Gamma$. Then the set of all elements which map v_k to v_1 is HT_k, a coset which has m elements, so the $S^rT_k(F)$ have v_1 as a vertex with an angle of θ_k. On the other hand, if a region $A(F)$ $(A\in\Gamma)$ has v_1 as a vertex, then $A^{-1}(v_1)\in F$, hence $A^{-1}(v_1)=v_i$ for some i, $1\le i\le t$. Thus $A\in HT_i$, and $A(F)$ has been included in the above description. So we have mt regions surrounding v_1. These regions are distinct, for if $S^rT_k(F)=S^qT_l(F)$, then $S^rT_k=S^qT_l$, and hence $r=q$ and $k=l$. We conclude then that

$$m(\theta_1+...+\theta_t)=2\pi. \qquad \square$$

We now consider the congruence of sides. Let s be a side of F, a Dirichlet region for a Fuchsian group Γ. If $T\in\Gamma-\{Id\}$ and $T(s)$ is a side of F, then s and $T(s)$ are called *congruent sides.* But $T(s)$ is also a side of $T(F)$ so that $T(s)\subseteq F\cap T(F)$. If a side of F has a fixed point of an elliptic element S of order 2 on it then S interchanges the two segments of this side. It is convenient to regard these two segments as distinct sides separated by a vertex. With this convention, it follows from Exercises 3.11 and 3.13 that for each side of F there exists another side of F congruent to it. There cannot be more than two sides in a congruent set. For, suppose that for some $T_1\in\Gamma$, $T_1(s)$ is also a side of F; then $T_1(s)=F\cap T_1(F)$. Thus $s=T_1^{-1}(F)\cap F=T^{-1}(F)\cap F$, so that $T_1^{-1}(F)\cap T^{-1}(F)\neq\emptyset$ which implies $T_1=T$. Thus the sides of F fall into congruent pairs. Hence if the number of sides of a Dirichlet region is finite, it is always even.

EXAMPLE A. The two vertical sides of the fundamental region for the modular group found in §3.2 are congruent via the transformation $z \rightarrow z+1$. The arc of the unit circle between ρ and $\rho+1$ (see Fig. 9) is the union of two sides: $[\rho, i]$ and $[i, \rho+1]$, congruent via the elliptic transformation of order 2, $z \rightarrow -1/z$.

THEOREM 3.5.4. *Let* $\{T_i\}$ *be the subset of* Γ *consisting of those elements which pair the sides of some fixed Dirichlet region* F. *Then* $\{T_i\}$ *is a set of generators for* Γ.

PROOF: Let Λ be the subgroup generated by the set $\{T_i\}$. We have to show that $\Lambda=\Gamma$. Suppose that $S_1 \in \Lambda$, and that $S_2(F)$ is adjacent to $S_1(F)$, i.e. they share a side. Then $S_1^{-1}S_2(F)$ is adjacent to F. Hence $S_1^{-1}S_2=T_k$ for some $T_k \in \{T_i\}$; and since $S_2=S_1T_k$ we conclude that $S_2 \in \Lambda$. Suppose now $S_3(F)$ intersects $S_1(F)$ in a vertex v. Then $S_1^{-1}S_3(F)$ intersects F in a vertex $u=S_1^{-1}v$. By Theorem 3.5.1, there can only be finitely many faces with vertex u, and F can be "connected" with $S_1^{-1}S_3(F)$ by a finite chain of faces in such a way that each two consecutive ones share a side. Hence we can apply the above argument repeatedly to show that $S_3 \in \Lambda$. Let $X=\bigcup_{S \in \Lambda} S(F)$, $Y=\bigcup_{S \in \Gamma-\Lambda} S(F)$. Then $X \cap Y=\emptyset$. Clearly $X \cup Y=\mathcal{H}$, so if we show that X and Y are closed subsets of $\mathcal{H}$, then as $\mathcal{H}$ is connected and $X \neq \emptyset$, we must have $X=\mathcal{H}$ and $Y=\emptyset$. This would show that $\Lambda=\Gamma$ and the result will follow.

We now show that any union $\bigcup V_j(F)$ of faces of the tessellation is closed. Suppose $\{z_i\}$ is an infinite sequence of points of $\bigcup V_j(F)$ which tends to some limit $z_0 \in \mathcal{H}$. Then $z_0 \in T(F)$ for some $T \in \Gamma$, and by Theorem 3.5.1, there exists a neighborhood N of z_0

intersecting only finitely many of the $V_j(F)$. Therefore, one face of this finite family, say $V_m(F)$, must contain a subsequence of $\{z_i\}$ tending to z_0. Since $V_m(F)$ is closed, $z_0 \in V_m(F) \subseteq \bigcup V_j(F)$. Thus $\bigcup V_j(F)$ is closed, and in particular X and Y are closed. □

EXAMPLE A. Theorem 3.5.4 implies that the modular group is generated by $z \to z+1$ and $z \to -1/z$.

3.6. Connection with Riemann surfaces and homogeneous spaces

Let Γ be a Fuchsian group acting on the upper half-plane $\mathcal{H}$ with $\mu(\Gamma\backslash\mathcal{H})<\infty$, and F be a fundamental region for this action. The group Γ induces a natural projection (continuous and open) π: $\mathcal{H} \to \Gamma\backslash\mathcal{H}$, and the points of $\Gamma\backslash\mathcal{H}$ are the Γ-orbits. The restriction of π to F identifies the congruent points of F that necessarily belong to its boundary ∂F, and makes $\Gamma\backslash F$ into an oriented surface with possibly some *marked points* (which correspond to the elliptic cycles of F) and *cusps* (which correspond to non-congruent vertices at infinity of F), also known as an *orbifold.* Its topological type is determined by the number of cusps and by its *genus*—the number of handles if we view the surface as a sphere with handles. If F is locally finite, the quotient space $\Gamma\backslash\mathcal{H}$ is homeomorphic to $\Gamma\backslash F$ [B, Thm. 9.2.4], hence by choosing F to be a Dirichlet region which is locally finite by Theorem 3.5.1, we can find the topological type of $\Gamma\backslash\mathcal{H}$. We have seen in §3.1 (Theorem 3.1.1) that the area of a fundamental region (with nice boundary) is, if finite, a numerical invariant of the group Γ. Since the area on the quotient space $\Gamma\backslash\mathcal{H}$ is induced by the hyperbolic area on $\mathcal{H}$, the *hyperbolic area of* $\Gamma\backslash\mathcal{H}$, denoted by $\mu(\Gamma\backslash\mathcal{H})$, is well defined and equal to $\mu(F)$ for any fundamental region F. If Γ has a compact Dirichlet region F, then by Exercise 3.10, F has finitely many sides, and the quotient space $\Gamma\backslash\mathcal{H}$

is compact. We shall see in §4.2 (Cor. 4.2.3) that if one Dirichlet region for Γ is compact then all Dirichlet regions are compact. If, in addition, Γ acts on $\mathcal{H}$ without fixed points, $\Gamma\backslash\mathcal{H}$ is a compact *Riemann surface*—a 1-dimensional complex manifold—and its fundamental group is isomorphic to Γ [Sp].

In many respects Fuchsian groups are similar to the lattices in $\mathbb{R}^n$ which are discrete groups of orientation-preserving Euclidean isometries. However, while the quotients of the latter are always compact surfaces homeomorphic to the torus, *the quotient of a Fuchsian group Γ acting on $\mathcal{H}$ without fixed points cannot be a torus.* For, since the fundamental group of a torus is isomorphic to $\mathbb{Z}\times\mathbb{Z}$, such a Γ must be isomorphic to $\mathbb{Z}\times\mathbb{Z}$ which contradicts Corollary 2.3.7. Equivalently, no compact Riemann surface of genus 1 has its universal covering space conformally equivalent to $\mathcal{H}$. We shall see, however, that *all orientable surfaces (compact or not) other than the sphere, torus, plane, or punctured plane are quotients of Fuchsian groups acting on $\mathcal{H}$ without fixed points* (see Corollary 4.3.3 and the remark at the end of §4.3).

Since Γ acts on $PSL(2,\mathbb{R})$ by left multiplication one can form the homogeneous space $\Gamma\backslash PSL(2,\mathbb{R})$. We have seen (Theorem 2.1.1) that $PSL(2,\mathbb{R})$ can be interpreted as the unit tangent bundle of the upper half-plane. It is easy to see (Exercise 3.14) that if F is a fundamental region for Γ in $\mathcal{H}$, SF is a fundamental region for Γ in $PSL(2,\mathbb{R})$. It also can be shown (see Exercise 3.15) that if Γ contains no elliptic elements, the homeomorphism described in Theorem 2.1.1 induces an homeomorphism of the corresponding quotient spaces. If Γ contains elliptic elements, an analogous result holds; however, the structure of the fibered bundle is violated in a finite number of marked points.

Since the fiber in $S(\Gamma\backslash\mathcal{H})$ over each point of $\Gamma\backslash\mathcal{H}$ is compact,

$\Gamma\backslash\mathcal{H}$ is compact if and only if $S(\Gamma\backslash\mathcal{H})$ is compact.

Now we give a sufficient condition for a fundamental region $\mathcal{F}=SF$ for Γ in $S\mathcal{H}$ to be compact.

THEOREM 3.6.1. *Suppose there exists a compact subset* $K\subset S\mathcal{H}$ *such that* $S\mathcal{H}=\Gamma\cdot K$, *i.e. any* $(z,\zeta)\in S\mathcal{H}$ *can be written as*

$$(z,\zeta)=T(k,\kappa),\quad T\in\Gamma,\ (k,\kappa)\in K.$$

Then any $\mathcal{F}=SF$, *where* $F=D_{z_0}(\Gamma)$ *is a Dirichlet region for* Γ *in* $\mathcal{H}$, *is compact.*

PROOF: Let l be the distance function on $S\mathcal{H}$ corresponding to the Riemannian metric $d\ell=\sqrt{ds^2+d\theta^2}$ (see §2.1). Suppose $\mathcal{F}$ is not compact. Then F is not compact, and hence is not bounded, i.e. there exists a sequence $\{z_n\}\in F$ such that $\rho(z_0,z_n)\to\infty$. Then for any (z_0,ζ_0) and $\{(z_n,\zeta_n)\}$ in $S\mathcal{H}$ we have

$$l((z_0,\zeta_0),(z_n,\zeta_n))\to\infty.$$

Since F is a Dirichlet region,

$$\rho(z_0,z_n)\leq\rho(T(z_0),z_n) \tag{3.6.1}$$

for all $T\in\Gamma$. We have $(z_n,\zeta_n)=T_n(k_n,\kappa_n)=(T_n(k_n),DT_n(\kappa_n))$. Plugging $T=T_n$ into (3.6.1), we get

$$\rho(z_0,z_n)\leq\rho(T_n(z_0),z_n)=\rho(T_n(z_0),T_n(k_n))=\rho(z_0,k_n).$$

But since K is compact, $\{l((z_0,\zeta_0),(k_n,\kappa_n))\}$ is bounded, hence $\{\rho(z_0,k_n)\}$ is bounded, and $\{\rho(z_0,z_n)\}$ is bounded, a contradiction. □

COROLLARY 3.6.2. *Suppose* Γ^1 *is a subgroup of* $SL(2,\mathbb{R})$ *containing* (-1_2), *and* $\Gamma=\Gamma^1/\{+1_2,-1_2\}$. *If there exists a compact subset* $K^1\subset SL(2,\mathbb{R})$ *such that* $SL(2,\mathbb{R})=\Gamma^1\cdot K^1$, *then* $\Gamma\backslash\mathcal{H}$ *is compact.* □

EXERCISES FOR CHAPTER 3

3.1. Let $z_1=x_1+iy_1$ and $z_2=x_2+iy_2$ be two points in $\mathcal{H}$ and let L be the perpendicular bisector of the geodesic segment $[z_1,z_2]$. Prove that for any z_1 and any compact subset K of $\mathbb{R}^2$, $L\cap K=\emptyset$ when $|z_2|$ is sufficiently large.

3.2. Show that the Dirichlet region can be described using the Euclidean metric as follows:

$$D_p(\Gamma)=\left\{z\in\mathcal{H} \mid \left|\frac{T(z)-p}{z-p}\right|\geq\frac{1}{|cz+d|} \text{ for all } T\in\Gamma\right\}.$$

3.3. Prove Theorem 3.3.3 for the unit disc model.

3.4 Suppose $T,S,R\in PSL(2,\mathbb{R})$ and $S=R^{-1}\circ T\circ R$. What happens to the isometric circles under this conjugation? Consider the same question for the group of orientation-preserving isometries (3.3.2) of the unit disc $\mathcal{U}$.

3.5. Prove that

(i) T is hyperbolic if and only if $I(T)$ and $I(T^{-1})$ do not intersect,

(ii) T is elliptic if and only if $I(T)$ and $I(T^{-1})$ intersect,

(iii) T is parabolic if and only if $I(T)$ and $I(T^{-1})$ are tangential.

3.6. Prove Theorem 3.3.4 for the upper half-plane model.

3.7. Give an alternative proof of Theorem 3.3.5 by showing that for any orientation-preserving isometry T of the unit disc the perpendicular bisector of $[0, T(0)]$ is the isometric circle $I(T^{-1})$.

3.8. Prove that a Fuchsian group is elementary if and only if its limit set consists of not more than two points.

3.9. Prove that if a fundamental region for a Fuchsian group F is locally finite, then for each point $z\in\mathcal{H}$ there is a compact neighborhood V and a finite set of elements $T_1, \dots, T_n\in\Gamma$ such that

(i) $z\in T_1(F)\cap \dots \cap T_n(F)$,

(ii) $V\subset T_1(F)\cup \dots \cup T_n(F)$,

(iii) $T(F) \cap V = \emptyset$ if $T \neq T_i$, $i=1, \ldots, n$

3.10. Show that the vertices of a Dirichlet region are *isolated*, that is every vertex of F has a neighborhood containing no other vertices of F. Deduce that a compact Dirichlet region has a finite number of vertices.

3.11. Prove that if F is a locally finite fundamental region, then for each $z \in \partial F$ there exists $Id \neq T \in \Gamma$ such that $T(z) \in \partial F$.

3.12. Show that ρ and $\rho+1$ are the only 2 elliptic points of order 3 and i is the only elliptic point of order 2 in the fundamental region F for $PSL(2,\mathbb{Z})$ in Figure 11.

3.13. Let F be a Dirichlet region for a Fuchsian group Γ and let s be a side of F. If $T \in \Gamma$ and T(s) is a side of F, prove that $F \cap T(F) = T(s)$.

3.14. Prove that if F is a fundamental region for a Fuchsian group Γ on $\mathcal{H}$, then SF is a fundamental region for Γ on $S\mathcal{H}$.

3.15. Prove that if Γ is a Fuchsian group without elliptic elements, then $S(\Gamma\backslash\mathcal{H})$ is homeomorphic to $\Gamma\backslash PSL(2,\mathbb{R})$.

4. GEOMETRY OF FUCHSIAN GROUPS

4.1. Geometrically finite Fuchsian groups

We saw in §3.5 that a Dirichlet region in $\mathcal{H}$ of a Fuchsian group Γ is bounded by a number of geodesic segments and possibly segments of the real axis. Here we shall be a little more precise. The bounding geodesic segments in $\mathcal{H}$ are called *sides* of F and form the *boundary of* F, ∂F; if an elliptic fixed point of order 2 belongs to a geodesic segment we include this point in the set of *vertices* of F and regard the two subsegments as two sides. Recall also that F is a closed subset of $\mathcal{H}$, but it may not be a closed subset of $\widetilde{\mathcal{H}}$. We call the closure of F in $\widetilde{\mathcal{H}}$ the *Euclidean closure of* F and denote it by $\widetilde{F}$. We also define the *Euclidean boundary of* F, $\partial_0 F$: $\partial_0 F = \widetilde{F} - F$ which obviously belongs to the set of *points at infinity*. $\partial_0 F$ may have uncountably many components, but there can only be countably many components of positive (Euclidean) length: we call these *free sides* of F; *vertices at infinity* of F (see §3.5) also belong to $\partial_0 F$.

DEFINITION. A Fuchsian group Γ is called *geometrically finite* if there exists a convex fundamental region for Γ with finitely many sides.

THEOREM 4.1.1. (Siegel's Theorem) *If* Γ *is such that* $\mu(\Gamma\backslash\mathcal{H})<\infty$ *then* Γ *is geometrically finite.*

PROOF [GP]: We shall prove that any Dirichlet region $F = D_p(\Gamma)$ has finitely many sides. Since the vertices of $D_p(\Gamma)$ are isolated (Exercise 3.10), any compact subset $K \subset \mathcal{H}$ contains only finitely many vertices.

This takes care of the case in which F is compact. Now suppose that F is not compact.

The main ingredient of the proof is an estimation of the angles ω at vertices of the region F. More precisely, we are going to prove that

$$\sum_{\omega}(\pi-\omega)\leq\mu(\mathrm{F})+2\pi, \tag{4.1.1}$$

where the sum is taken over all vertices of F lying in $\mathcal{H}$. We first notice that F is a star-like generalized polygon, and that the boundary of F, ∂F, is not necessarily connected.

Let us connect all vertices of F with the point p by geodesics and consider the triangles thus obtained. Let ... A_m, A_{m+1}, ... , A_n, ... be a connected set of geodesic segments in ∂F with vertices ... a_m, a_{m+1}, ... , a_{n+1}, ... (Fig. 12). We assume that this set is unbounded in both directions. We denote the triangle with the side A_k by Δ_k, its angles by α_k, β_k, γ_k, and the angle between A_k and A_{k+1} by ω_k; thus we have

$$\omega_k=\beta_k+\gamma_{k+1}.$$

By the Gauss-Bonnet formula (Theorem 1.4.2) we have

$$\mu(\Delta_k)=\pi-\alpha_k-\beta_k-\gamma_k.$$

Thus

$$\sum_{k=m}^{n}\alpha_k+\sum_{k=m}^{n}\mu(\Delta_k)=\pi-\gamma_m-\beta_n+\sum_{k=m}^{n-1}(\pi-\omega_k). \tag{4.1.2}$$

The left-hand side of this equality is bounded since $\sum\alpha_k\leq2\pi$ and $\sum\mu(\Delta_k)\leq\mu(\mathrm{F})$, hence the right-hand side is also bounded. It follows that $\sum(\pi-\omega_k)$ converges, and the following limits exist:

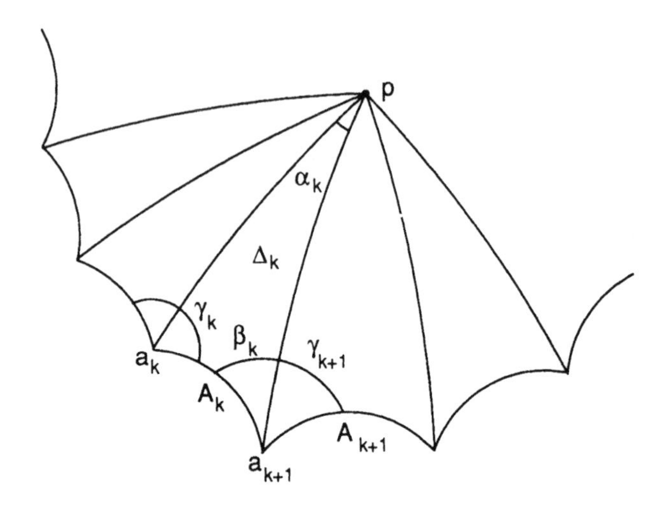

Fig. 12

$$\lim_{m\to-\infty}\gamma_m=\gamma_{-\infty},\quad \lim_{n\to\infty}\beta_n=\beta_\infty.$$

Let us show that

$$\pi-\gamma_{-\infty}-\beta_\infty\geq 0. \tag{4.1.3}$$

Since only finitely many segments $\{A_k\}$ may be a bounded distance from the point p, we have $a_k\to\infty$ as $k\to\infty$. Thus $\rho(p,a_{k+1})>\rho(p,a_k)$ for infinitely many values of k, and for these values, as follows, for instance, from the Sine Rule (Theorem 1.5.2(i)), we have $\gamma_k>\beta_k$. On the other hand, $\beta_k+\gamma_k\leq\pi$ and thus $\beta_k\leq\pi/2$. Therefore $\beta_\infty\leq\pi/2$. Similarly, $\gamma_{-\infty}\leq\pi/2$, and (4.1.3) follows.

Let $m\to-\infty$, $n\to\infty$. Taking into account (4.1.3) we obtain from (4.1.2) a limit inequality

$$\sum_{k=-\infty}^{\infty} \alpha_k + \sum_{k=-\infty}^{\infty} \mu(\Delta_k) \geq \sum_{k=-\infty}^{\infty} (\pi - \omega_k). \tag{4.1.4}$$

The inequality is obtained under the assumption that the connected set of segments $\{A_k\}$ is unbounded in both directions. Similar arguments apply in other cases when the connected set of segments is bounded at least in one direction. Adding up all these inequalities, we obtain a desired estimate

$$2\pi + \mu(F) \geq \sum_{\omega} (\pi - \omega), \tag{4.1.5}$$

where the sum is taken over all vertices of F which lie a finite hyperbolic distance from the point p, i.e. in $\mathcal{H}$.

Now we are going to prove, using this estimate, that the number of vertices which lie a finite distance from the point p is finite. Let a be a vertex and $a^{(1)}=a$, $a^{(2)}$, ... , $a^{(n)}$ all vertices congruent to a. If we denote the angle at vertex $a^{(i)}$ by $\omega^{(i)}$, we have by Theorem 3.5.3 :

$$\omega^{(1)} + \omega^{(2)} + \dots + \omega^{(n)} = 2\pi, \tag{4.1.6}$$

if a is not a fixed point for any $T \in \Gamma - \{Id\}$; and

$$\omega^{(1)} + \omega^{(2)} + \dots + \omega^{(n)} = 2\pi/m, \tag{4.1.7}$$

if a is a fixed point of order m. Since $\omega^{(i)} < \pi$ for each cycle of the type (4.1.6), we have $n \geq 3$, and hence

$$\sum_{i=1}^{n} (\pi - \omega^{(i)}) = (n-2)\pi > \pi. \tag{4.1.8}$$

Comparing (4.1.8) with (4.1.5) we conclude that the number of cycles where a is not a fixed point for any $T \in \Gamma - \{Id\}$ is finite. For each cycle of the type (4.1.7) we have

$$\sum_{i=1}^{n} (\pi - \omega^{(i)}) = (n - \tfrac{2}{m})\pi > \tfrac{\pi}{3}. \tag{4.1.9}$$

Comparing (4.1.9) with (4.1.5) we conclude that the number of elliptic cycles of order$\geq$3 is finite. Finally, any elliptic fixed point of order 2 belongs to a segment of ∂F between two vertices which are not elliptic points of order 2, hence we see that the number of elliptic cycles of order 2 is also finite. Thus we have proved that there are only finitely many vertices a finite distance from the point p.

It remains to show that the number of vertices at infinity is also finite. Let us take any N vertices at infinity: $B_1, \dots, B_N$. It is obvious that there exists a hyperbolic polygon F_1 bounded by a finite number of geodesics and contained inside F such that its vertices at infinity are $B_1, \dots, B_N$. An argument similar to that in the proof of (4.1.4) shows that the hyperbolic area of F_1 satisfies the following equation:

$$\sum_{\omega} (\pi - \omega) = 2\pi + \mu(F_1),$$

where ω are the angles at the vertices of F_1, and the sum is taken over all vertices of F_1. Since $\omega=0$ for all vertices at infinity, we have

$$\pi N \leq 2\pi + \mu(F_1) \leq 2\pi + \mu(F).$$

Thus N is bounded from above, and the theorem follows. □

4.2. Cocompact Fuchsian groups

DEFINITION. A Fuchsian group is called *cocompact* if the quotient-space $\Gamma\backslash\mathcal{H}$ is compact.

The following results reveal the relationship between cocompactness of Γ and the absence of parabolic elements in Γ.

THEOREM 4.2.1. *If a Fuchsian group Γ has a compact Dirichlet region, then Γ contains no parabolic elements.*

PROOF: Let F be a compact Dirichlet region for Γ, and

$$\eta(z)=\inf\{\rho(z,T(z)) \mid T\in\Gamma-\{Id\},\ T \text{ not elliptic}\}.$$

Since the Γ-orbit of each $z\in\mathcal{H}$ is a discrete set (Cor. 2.2.7) and T(z) is continuous, $\eta(z)$ is a continuous function of z and $\eta(z)>0$. Therefore, as F is compact, $\eta=\inf\{\eta(z) \mid z\in F\}$ is attained and $\eta>0$. If $z\in\mathcal{H}$, there exists $S\in\Gamma$ such that $w=S(z)\in F$. Hence, if $T_0\in\Gamma-\{Id\}$ is not elliptic,

$$\rho(z,T_0(z))=\rho(S(z),S(T_0(z))=\rho(w,ST_0S^{-1}(w))\geq\eta,$$

and therefore

$$\inf\{\rho(z,T_0(z)) \mid z\in\mathcal{H},\ T_0 \text{ not elliptic}\}=\eta>0.$$

Now suppose that Γ contains a parabolic element T_1. If for some $R\in PSL(2,\mathbb{R})$, $\Gamma_1=R\Gamma R^{-1}$ then R(F) will be a compact fundamental region for Γ_1. Thus by conjugating Γ in $PSL(2,\mathbb{R})$ we may assume that $T_1(z)$ or $T_1^{-1}(z)$ is the transformation $z\rightarrow z+1$. However, by Theorem 1.2.6(iii), $\rho(z,z+1)\rightarrow 0$ as $Im(z)\rightarrow\infty$, a contradiction. □

THEOREM 4.2.2.

(i) If Γ has a non-compact Dirichlet region then the quotient space $\Gamma\backslash\mathcal{H}$ is not compact.

(ii) If a Dirichlet region $F=D_p(\Gamma)$ *for a Fuchsian group* Γ *has finite hyperbolic area but is not compact, then it has at least one vertex at infinity.*

PROOF: Let $F=D_p(\Gamma)$ be a non-compact Dirichlet region for Γ. We consider all oriented geodesic rays from the point p; each geodesic ray is uniquely determined by its direction l at the point p. Since F is a hyperbolically convex region, a geodesic ray in the direction l either intersects ∂F in a unique point or the whole geodesic ray lies inside F. Hence we can define a function $\tau(l)$ to be the length of a geodesic segment in the direction l inside F, $\tau(l)$ being equal to ∞ in the latter case. Obviously, $\tau(l)$ is a continuous function of l at the points where $\tau(l)<\infty$. Therefore if $\tau(l)<\infty$ for all l, the function $\tau(l)$ is bounded; hence the region F is compact. Thus if F is not compact, there are some directions l for which $\tau(l)=\infty$. After the identification of the congruent points of ∂F, we obtain a non-compact orbifold $\Gamma\backslash\mathcal{H}$ and (i) follows. To prove (ii), let us consider one such direction l_0. The intersection of the geodesic ray from p in the direction l_0 with the set of points at infinity belongs to $\partial_0 F$, the Euclidean boundary of F. By Theorem 4.1.1, F is geometrically finite, hence $\partial_0 F$ consists of finitely many free sides and vertices at infinity. Since $\mu(F)<\infty$, it is easy to see that $\partial_0 F$ cannot contain any free sides. Therefore this intersection is a vertex at infinity, and (ii) follows. □

COROLLARY 4.2.3. *The quotient space of a Fuchsian group* Γ, $\Gamma\backslash\mathcal{H}$ *is compact if and only if any Dirichlet region for* Γ *is compact.* □

Let $p\in\mathcal{H}$ and $z(t)$, $0\leq t<\infty$, be a geodesic ray from the point p. Let $B_t(p)$ be a hyperbolic circle centered at $z(t)$ and passing through the point p. Exercise 4.2 asserts that the limit of $B_t(p)$, as $t\to\infty$, exists. It is a Euclidean circle passing through p and through the end of the geodesic z(t) corresponding to t=∞, and orthogonal to the geodesic z(t). Since the geodesic ray through p is determined by its direction l, the limiting circle depends on p and l. It is called a *horocycle* (see Fig. 13) and is denoted by $\omega(p, l)$. Notice that horocycles are not hyperbolic circles, however they may be considered as circles of infinite radius. Let s$\in\mathbb{R}$. A Euclidean circle through s tangent to the real line is a horocycle for any geodesic in $\mathcal{H}$ represented by a Euclidean semicircle passing through s (and orthogonal to the real axis, of course). (see Fig. 14.) If s=∞, the geodesics passing through ∞ are represented by vertical straight lines, and horocycles are Euclidean straight lines parallel to the real axis (see Fig. 15). We shall denote a horocycle through a point s at infinity by $\omega(s)$. We have a whole family of horocycles $\omega(s)$ through a given point s.

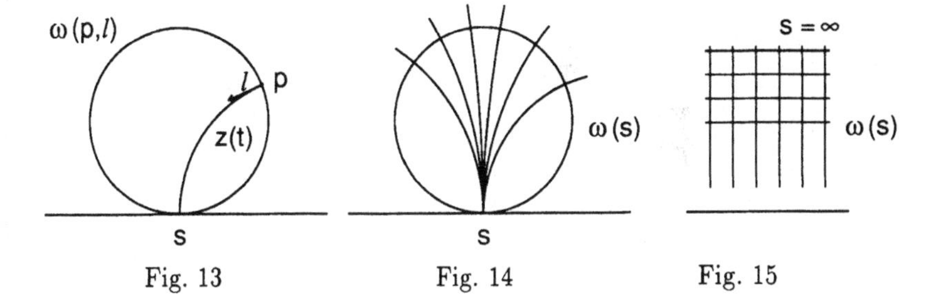

Fig. 13 Fig. 14 Fig. 15

THEOREM 4.2.4. *Let* S *be a transformation in* PSL(2,$\mathbb{R}$) *fixing a point* s$\in\mathbb{R}$. *Then* S *is parabolic if and only if for each horocycle through* s, $\omega(s)$, *we have* $S(\omega(s))=\omega(s)$.

PROOF: Suppose first that S is parabolic, and R$\in$PSL(2,$\mathbb{R}$) is such that R(s)=∞. Then $S_0=R\circ S\circ R^{-1}$ is a parabolic transformation fixing ∞, and therefore S(z)=z+h, h$\in\mathbb{R}$. Since S is a Euclidean translation, it maps each horizontal line to itself. Since a linear fractional transformation maps circles and straight lines into circles and straight lines and preserves angles, we conclude that horocycles are mapped into horocycles. Thus $S(\omega(s))=\omega(s)$.

Conversely, suppose S maps each horocycle $\omega(s)$ onto itself. Making the same conjugation as above, we move the fixed point s to ∞. Then $S(z)=az+b$. The condition that each horizontal line is mapped into itself implies that $a=1$. Hence S is a parabolic element□

THEOREM 4.2.5. *Suppose* Γ *has a non-compact Dirichlet region* $F=D_p(\Gamma)$ *with* $\mu(F)<\infty$. *Then*
(i) each vertex of F *at infinity is a parabolic fixed point for some* $T\in\Gamma$.
(ii) If ξ *is a fixed point of some parabolic element in* Γ, *then there exists* $T\in\Gamma$ s.t. $T(\xi)\in\partial_0(F)$.

PROOF of (i): Let b be a vertex of F at infinity. Let us consider all images S(F), $S\in\Gamma$, which have the point b as a vertex. Obviously, there are infinitely many of them. Let $b^{(1)}=b$, $b^{(2)}$, ... , $b^{(n)}$ be all vertices of F congruent to b:

$$b^{(k)}=T_k(b) \quad (k=1, \dots , n).$$

We know from Theorem 4.1.1 that the number of such vertices is finite. Any image of F which has the point b as a vertex has a form

$$TT_i^{-1}(F) \quad (i=1, \dots , n),$$

where T is any element of Γ which fixes the point b. Since there are infinitely many such images, and since T_i is only taken from a finite set of elements, we conclude that there are infinitely many elements

$T \in \Gamma$ fixing b.

We shall show now that any such element T is a parabolic element. Suppose T is not parabolic. Let us consider a geodesic z(t), $0 \le t \le \infty$, parametrized by its length, connecting the points p and b ($z(0)=p$, $z(\infty)=b$). (See Fig. 16.) Since F is a Dirichlet region the whole geodesic lies inside F and

$$\rho(p,z(t)) < \rho(T(p),z(t)), \quad 0 \le t < \infty. \tag{4.2.1}$$

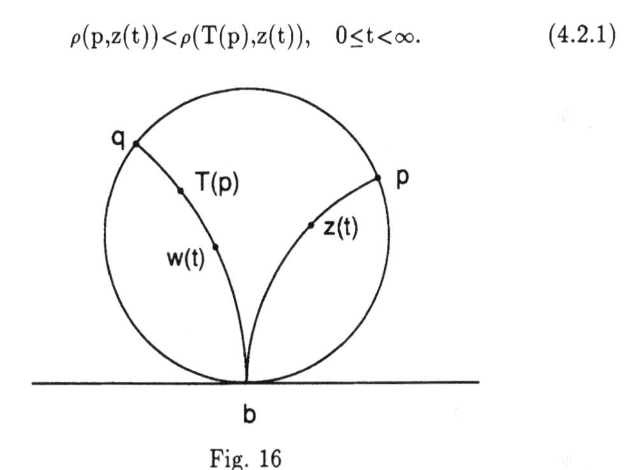

Fig. 16

Consider a horocycle $\omega(b)$ containing the point p. Since by our assumption T is not a parabolic transformation, T(p) does not belong to $\omega(b)$. Then by Exercise 4.3 either T(p) or $T^{-1}(p)$ lies inside $\omega(b)$. We may assume then that T(p) lies inside $\omega(b)$. Let w(t) be a geodesic passing through T(p) and b. Let q be a second point of intersection of $\omega(b)$ and w(t); we choose the parametrization of w(t) by its length such that $w(0)=q$. We notice first that $\rho(z(t),w(t)) \to 0$ as $t \to \infty$. In order to see this, we conjugate Γ so that its action on $\mathcal{H}$ gives: $b=\infty$, $z(t)=a+it$, $w(t)=c+it$ $(t \ge t_0 > 0)$. Then using Theorem 1.2.8(iii), we obtain

$$\sinh\left[\tfrac{1}{2}\rho(z(t),w(t))\right]=\frac{|a-c|}{2t}\to 0 \text{ as } t\to 0,$$

and the claim follows. We have

$$t=\rho(p,z(t))=\rho(q,w(t))=\rho(q,T(p))+\rho(T(p),w(t))$$

$$\geq\rho(q,T(p))+\rho(T(p),z(t))-\rho(z(t),w(t)),$$

and hence for sufficiently large t, we have

$$\rho(p,z(t))>\rho(T(p),z(t)),$$

a contradiction with (4.2.1).

PROOF of (ii): See Exercise 4.4. □

We leave the proof of the following Corollary (Exercise 4.5).

COROLLARY 4.2.6. *There is a one-to-one correspondence between non-congruent vertices at infinity of a Dirichlet fundamental region for a non-cocompact Fuchsian group* Γ *with* $\mu(\Gamma\backslash\mathcal{H})<\infty$ *and conjugacy classes of maximal parabolic subgroups of* Γ.

The following result is a direct consequence of Theorems 4.1.1, 4.2.2, and 4.2.5.

COROLLARY 4.2.7. *A Fuchsian group* Γ *is cocompact if and only if* $\mu(\Gamma\backslash\mathcal{H})<\infty$ *and* Γ *contains no parabolic elements.*

4.3. The signature of a Fuchsian group

We now assume that Γ has a compact fundamental region F. By Exercise 3.10, F has finitely many sides, and hence finitely many vertices, finitely many elliptic cycles, and by Theorem 3.5.2, a finite number of periods, say $m_1, \dots, m_r$. As we have seen in §3.6 the quotient space $\Gamma\backslash\mathcal{H}$ is an orbifold, i.e a compact, oriented surface of genus g with exactly r marked points. In this case we say that Γ has *signature* $(g; m_1, m_2, \ldots, m_r)$.

THEOREM 4.3.1. *Let* Γ *have signature* $(g; m_1, \ldots, m_r)$. *Then*

$$\mu(\Gamma\backslash\mathcal{H})=2\pi[(2g-2)+\sum_{i=1}^{r}\left(1-\frac{1}{m_i}\right)].$$

PROOF: The area of the quotient space was defined in the beginning of §3.6: $\mu(\Gamma\backslash\mathcal{H})=\mu(F)$ where F is a Dirichlet region. By Theorem 3.5.2 F has r elliptic cycles of vertices. (As described in §3.6 we include the interior point of a side fixed by an elliptic element of order 2 as a vertex whose angle is π, and then regard this side as being composed of two sides separated by this vertex.) By Theorem 3.5.3, the sum of angles at all elliptic vertices is $\sum_{i=1}^{r}\frac{2\pi}{m_i}$. Suppose there exist s other cycles of vertices. Since the order of the stabilizers of these vertices is equal to 1, the sum of angles at all these vertices is equal to $2\pi s$. Thus the sum of all angles of F is equal to

$$2\pi[\left(\sum_{i=1}^{r}\frac{1}{m_i}\right)+s].$$

The sides of F are matched up by elements of Γ. If we identify those matched sides, we obtain an orbifold of genus g. If F has n such sets of identified sides, we obtain a decomposition of $\Gamma\backslash\mathcal{H}$ into (r+s) vertices, n edges, and 1 simply connected face. By the Euler formula,

$$2-2g=(r+s)-n+1.$$

Exercise 4.6 gives a formula for the hyperbolic area of a hyperbolic polygon. Using it, we obtain

$$\mu(F)=(2n-2)\pi-2\pi[\left(\sum_{i=1}^{r}\frac{1}{m_i}\right)+s]=2\pi[(2g-2)+\sum_{i=1}^{r}\left(1-\frac{1}{m_i}\right)]. \qquad \square$$

It is quite surprising that the converse to Theorem 4.3.1 is also true, i.e. that there exists a Fuchsian group with a given signature. This first appeared in Poincaré's paper on Fuchsian groups [P], but the rigorous proof was given much later by Maskit [Ma]. Below we give a construction of a fundamental polygon for a Fuchsian group of the given signature that can be found in [JS].

THEOREM 4.3.2. (Poincaré's Theorem) *If* $g\geq 0$, $r\geq 0$, $m_i\geq 2$ $(1\leq i\leq r)$ *are integers and if*

$$(2g-2)+\sum_{i=1}^{r}\left(1-\frac{1}{m_i}\right)>0,$$

then there exists a Fuchsian group with signature $(g;\, m_1,\, \dots\, ,\, m_r)$.

SKETCH OF PROOF: We shall use the unit disc model $\mathcal{U}$ of hyperbolic geometry. From the center of $\mathcal{U}$ draw $(4g+r)$ radii at equal angles. Let $0<t<1$ and choose points at Euclidean distance t from the center on each radius. We join successive points by geodesics to get a regular hyperbolic polygon $M(t)$. On the first r sides of $M(t)$, we construct r external isosceles hyperbolic triangles such that the angles between the equal sides of the triangles are $2\pi/m_1, \dots, 2\pi/m_r$ (it is possible to do by Exercise 4.8; if $m_i=2$, the corresponding triangle will degenerate). In Figure 17 we have $g=2$, $r=4$, $m_i>2$ for $i=1, 2, 3$, and $m_4=2$. The union of $M(t)$ with these triangles is a star-like hyperbolic polygon $N(t)$ with $4g+2r$ sides. Label these sides $\alpha_1,\beta_1',\alpha_1',\beta_1, \dots, \alpha_g,\beta_g',\alpha_g',\beta_g, \xi_1,\xi_1', \dots, \xi_r,\xi_r'$, and orient them as indicated in Figure 17. If $t\to 0$, $\mu(N(t))\to 0$. Using Exercise 4.6 we obtain that $\mu(N(t))\to(4g+2r-2)\pi-\sum_{i=1}^{r}2\pi/m_i=$

$2\pi[(2g-1)+\sum_{i=1}^{r}\left(1-\frac{1}{m_i}\right)]$ as $t\to 1$. Hence, by continuity of $\mu(N(t))$, for some t_0 between 0 and 1, the hyperbolic area of $N(t_0)$ is exactly

$$2\pi[(2g-2)+\sum_{i=1}^{r}\left(1-\frac{1}{m_i}\right)].$$

By construction, α_i and α_i' have the same hyperbolic length as do β_j and β_j', and ξ_k and ξ_k'. By Exercise 1.1, for each pair of geodesics there exists an orientation-preserving isometry of $\mathcal{U}$ which maps one to the other. Hence there exist orientation-preserving hyperbolic

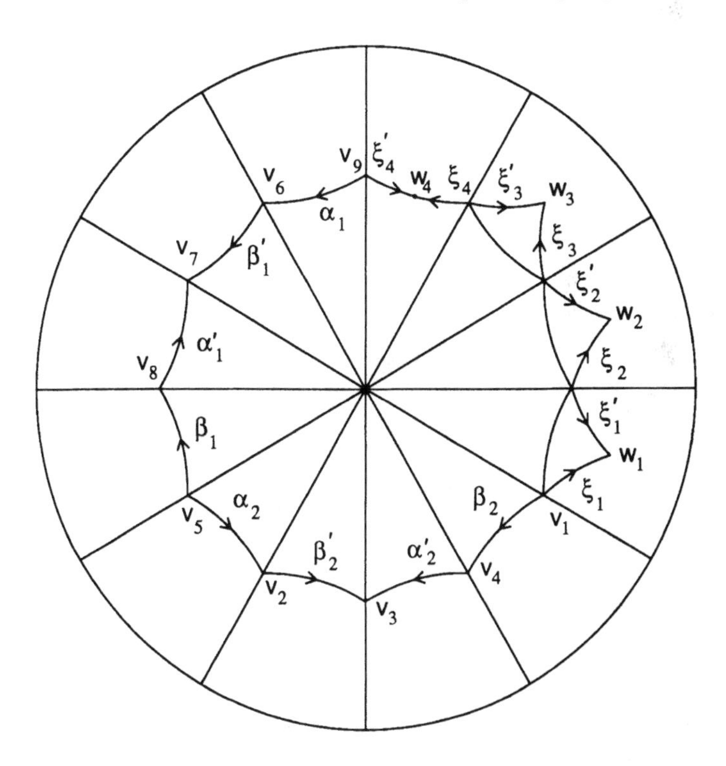

Fig. 17

isometries A_i, B_j, X_k (i,j=1, ..., g; k=1, ..., r) such that

$$A_i(\alpha_i')=\alpha_i,\ \ B_j(\beta_j')=\beta_j,\ \ X_k(\xi_k')=\xi_k.$$

Now we compute the congruence classes of vertices. Start by calling the vertex at beginning of β_g, v_1. It is congruent (via B_g^{-1}) to the beginning of β_g' which we will call v_2 and which is also the end of α_g. This is congruent, in its turn (via A_g^{-1}), to the end of α_g' which we will call v_3. The vertex v_3 is also the end of β_g' which is congruent (via B_g) to the end of β_g which we will call v_4. The vertex v_4 is also the beginning of α_g' which is congruent (via A_g) to the beginning of α_g which is, in its turn, a beginning of β_{g-1} (see Fig. 17). Proceeding with this process, we see that all the vertices of the original polygon $M(t_0)$ form a congruent set. Moreover, we conclude that

$$X_1X_2...X_rA_1B_1A_1^{-1}B_1^{-1}...A_gB_gA_g^{-1}B_g^{-1}(v_1)=v_1. \tag{4.3.1}$$

The other r vertices w_1, ... , w_r form r congruent sets each with just one element.

Let the sum of angles at the congruent set of vertices v_1, ... , v_{4g+r} be equal to Σ. As the hyperbolic area of $N(t_0)$ is $2\pi[(2g-2)+\sum_{i=1}^{r}\left(1-\frac{1}{m_i}\right)]$, by the Gauss-Bonnet formula (Exercise 4.6), we have

$$\Sigma=2\pi. \tag{4.3.2}$$

Now let Γ be the group generated by

$$\left\{A_i,\ B_j,\ X_k\ |i,j=1,\ ...,\ g;\ k=1,\ ...,\ r\right\}. \tag{4.3.3}$$

The necessary conditions of Theorem 3.5.3 are satisfied: the sum of the angles at v_1, v_2, ... , v_{4g+r} is equal to 2π and the angle at w_k is

$2\pi/m_k$ for each $k=1, \ldots, r$. It can be shown that the Γ-images of $N(t_0)$ cover $\mathcal{U}$ without overlap, so that $N(t_0)$ is a fundamental region for Γ. Thus the Γ-orbit of each point of $\mathcal{U}$ is a discrete set, and so by Corollary 2.2.7, Γ is a properly discontinuous group of hyperbolic isometries of $\mathcal{U}$, and if we transfer back to $\mathcal{H}$ we get a Fuchsian group.

The quotient space $\Gamma\backslash\mathcal{U}$ is decomposed into $(r+1)$ vertices (corresponding to the $(r+1)$ congruent sets of vertices of $N(t_0)$), $(2g+r)$ edges, and 1 simply connected face. By the Euler formula, its genus h satisfies

$$2-2h=(r+1)-(2g+r)+1=2-2g,$$

and therefore $h=g$. There are r elliptic cycles, namely $\{w_1\}, \ldots, \{w_r\}$, and their stabilizers have orders $m_1, \ldots, m_r$. Hence Γ has signature $(g; m_1, \ldots, m_r)$. □

If $r=0$, we obtain the following important result.

COROLLARY 4.3.3. *For any integer* $g>1$ *there exists a Fuchsian group acting on* $\mathcal{H}$ *without fixed points such that* $\Gamma\backslash\mathcal{H}$ *has genus* g.

EXAMPLE C. *Construction of a Fuchsian group with signature* $(2;-)$. Since $r=0$, a fundamental region is a regular hyperbolic octagon F_8 (see Fig. 18) of hyperbolic area 4π. We call this group Γ_8. Following the proof of Theorem 4.3.2, we suppose that t_0 is chosen such that $\mu(N(t_0))=4\pi$. Then the area of each of the 8 equalateral hyperbolic triangles is equal to $\frac{\pi}{2}$; and since the angle at the origin is equal $\frac{\pi}{4}$, by the Gauss-Bonnet formula (Theorem 1.4.2) the two other angles are equal to $\frac{\pi}{8}$. By Theorem 4.3.2, the group Γ_8 is generated by 4 hyperbolic elements, A_1, A_2, B_1, and B_2, that identify the sides of F_8 as shown in Figure 18. Since all eight sides of F_8 are arcs of circles of the same Euclidean radius of equal Euclidean length, the sides identified by a generator must be isometric circles of this generator and its inverse. This allows us to use elementary geometry to explicitely write down those generators. Let

$$A_2 = \begin{bmatrix} a & c \\ \bar{c} & \bar{a} \end{bmatrix}, \qquad (4.3.4)$$

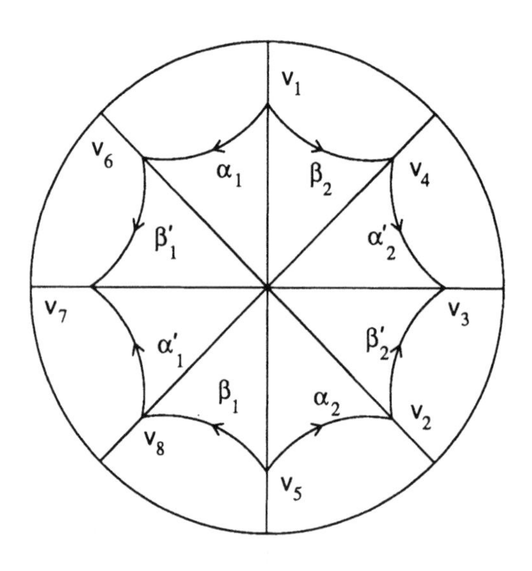

Fig. 18

then the isometric circle $I(A_2)$ is given by the equation $|\bar{c}z+\bar{a}|=1$.

By Theorem 3.3.4, A_2 maps $I(A_2)$ onto $I(A_2{}^{-1})$ in such a way that the center of $I(A_2)$, $-\frac{\bar{a}}{\bar{c}}$, is mapped onto the center of $I(A_2{}^{-1})$, $\frac{a}{\bar{c}}$. But from Figure, 18 we see that $\frac{ia}{\bar{c}}=-\frac{\bar{a}}{\bar{c}}$, which implies $a=\pm|a|(\frac{1}{\sqrt{2}}+i\frac{1}{\sqrt{2}})$. Let the radius of $I(A_2)=R$, and the distance of the center of $I(A_2)$ from the origin be d. By elementary geometric arguments, we have $d=R(1+\sqrt{2})$. On the other hand, $|c|=\frac{1}{R}$, and $d=\frac{|a|}{|c|}=R|a|$, hence $|a|=1+\sqrt{2}$; and since $|a|^2-|c|^2=1$, we have $|c|=\sqrt{2+2\sqrt{2}}$. Now let us choose the + sign in the expression for a, i.e. $\mathrm{Arg}(a)=\frac{\pi}{4}$. Since $\mathrm{Arg}(-\frac{\bar{a}}{\bar{c}})=\frac{\pi}{8}$, we obtain $\mathrm{Arg}(c)=-\frac{5\pi}{8}$. Using the formulas $\cos\frac{5\pi}{8}=-\frac{\sqrt{2-\sqrt{2}}}{2}$ and $\sin\frac{5\pi}{8}=\frac{\sqrt{2+\sqrt{2}}}{2}$, we obtain the expressions for the numbers a and c in (4.3.4):

$$a=\frac{2+\sqrt{2}}{2}(1+i),\ c=-\frac{\sqrt[4]{2}}{2}\left(\sqrt{2}+i(2+\sqrt{2})\right).$$

Other generators of the group Γ_8 can also be expressed in terms of parameters a and c as follows: $A_1=\begin{bmatrix} a & -c \\ -\bar{c} & \bar{a} \end{bmatrix}$, $B_1=\begin{bmatrix} \bar{a} & -\bar{c} \\ -c & a \end{bmatrix}$, $B_2=\begin{bmatrix} \bar{a} & \bar{c} \\ c & a \end{bmatrix}$.

Let $R: \mathcal{H} \to \mathcal{U}$ be a map given by $R(z)=\frac{zi+1}{z+i}$, see (1.2.2). Then $\Gamma=R^{-1}\Gamma_8 R$ be a subgroup of $PSL(2,\mathbb{R})$ whose generators are:

$$A_2=\begin{bmatrix} \mathrm{Re}(a)+\mathrm{Im}(c) & \mathrm{Im}(a)+\mathrm{Re}(c) \\ -(\mathrm{Im}(a)-\mathrm{Re}(c)) & \mathrm{Re}(a)-\mathrm{Im}(c) \end{bmatrix},$$

$$A_1=\begin{bmatrix} \mathrm{Re}(a)-\mathrm{Im}(c) & \mathrm{Im}(a)-\mathrm{Re}(c) \\ -(\mathrm{Im}(a)+\mathrm{Re}(c)) & \mathrm{Re}(a)+\mathrm{Im}(c) \end{bmatrix},$$

$$B_1=\begin{bmatrix} \mathrm{Re}(a)+\mathrm{Im}(c) & -\mathrm{Im}(a)-\mathrm{Re}(c) \\ \mathrm{Im}(a)-\mathrm{Re}(c) & -\mathrm{Re}(a)-\mathrm{Im}(c) \end{bmatrix},$$

$$B_2 = \begin{bmatrix} \mathrm{Re}(a) - \mathrm{Im}(c) & -\mathrm{Im}(a) + \mathrm{Re}(c) \\ \mathrm{Im}(a) + \mathrm{Re}(c) & \mathrm{Re}(a) + \mathrm{Im}(c) \end{bmatrix}.$$

This group will be considered in more detail in Chapter 5.

From the construction in the proof of Poincaré's Theorem, we can also extract information about relations between the generators of the group Γ. Since X_k fixes the point w_k of order m_k, for $k=1, \ldots, r$, we have

$$X_1^{m_1} = \ldots = X^{m_r} = \mathrm{Id}.$$

From (4.3.1) and the fact that the stabilizer of v_1 is only the identity element (compare with (4.3.2)), we obtain one more relation (notice that multiplication here is from right to left):

$$X_1 X_2 \ldots X_r A_1 B_1 A_1^{-1} B_1^{-1} \ldots A_g B_g A_g^{-1} B_g^{-1} = \mathrm{Id}.$$

Thus the presentation of a group Γ with signature $(g; m_1, \ldots, m_r)$ is

$$\langle A_1, B_1, A_2, B_2, \ldots, A_g, B_g, X_1, \ldots, X_r \mid X_1^{m_1} = \ldots = X^{m_r}$$
$$= X_1 X_2 \ldots X_r A_1 B_1 A_1^{-1} B_1^{-1} \ldots A_g B_g A_g^{-1} B_g^{-1} = \mathrm{Id} \rangle.$$

If

$$2\pi[(2g-2) + \sum_{i=1}^{r} \left(1 - \frac{1}{m_i}\right)] \leq 0,$$

there does not exist a Fuchsian group with signature $(g; m_1, \ldots, m_r)$. By a simple arithmetic calculation, there are only finitely many such signatures. For example, there is no Fuchsian group with signature $(1; -)$. This gives an alternative proof that no Fuchsian group without elliptic elements represents a compact Riemann surface of genus 1 (see the beginning of §3.6).

Suppose now that Γ has r conjugacy classes of maximal elliptic cyclic subgroups of orders $m_1, \ldots, m_r$, s conjugacy classes of maximal parabolic cyclic subgroups, and $\Gamma\backslash\mathcal{H}$ has genus g. Then we say that Γ has signature

$$(g;\ m_1,\ \dots,\ m_r;\ s). \tag{4.3.4}$$

The hyperbolic area of the quotient space can be computed in terms of signature (see Exercise 4.11). It also can be shown (Exrcise 4.10) that if $s>0$, $\mu(\Gamma\backslash\mathcal{H})\geq\pi/3$, the minimum being attained for the modular group which has signature (0; 2, 3; 1).

If the expression for the area in Exercise 4.11 is positive, then one can show, by a method similar to that of proof of Theorem 4.3.2, that a group Γ with signature (4.3.4) exists. (One needs s of the isosceles triangles to have vertices on Σ, the angle at these vertices being 0). The algebraic structure of the group Γ is determined by its signature, and a group with signature (4.3.4) has the presentation

$$<A_1,B_1,A_2,B_2,\ \dots\ ,A_g,B_g,\ X_1,\ \dots\ ,\ X_r,\ P_1,\ \dots\ ,\ P_s \mid {X_1}^{m_1}=\dots=X^{m_r}$$
$$=P_1P_2\dots P_sX_1X_2\dots X_rA_1B_1{A_1}^{-1}{B_1}^{-1}\dots A_gB_g{A_g}^{-1}{B_g}^{-1}=Id>,$$

where A_i, B_j, X_k are as in (4.3.3), and P_t $(t=1,\ \dots\ ,s)$ are the orientation-preserving isometries fixing vertices on Σ.

A Fuchsian group with signature $(0;\ m_1,\ \dots\ ,\ m_r;\ s)$ such that $r+s=3$ is called a *triangle group*; we shall also write $(0;\ m_1,m_2,m_3)$ for its signature slightly abusing our notations and allowing $m_i=\infty$. By Exercise 4.11 a triangle group may exist only if $\frac{1}{m_1}+\frac{1}{m_2}+\frac{1}{m_3}<1$.

4.4. Fuchsian groups generated by reflections

In this section we give a method for constructing triangle Fuchsian groups with given signature.

DEFINITION. Let Q be a geodesic in $\mathcal{H}$. A *hyperbolic reflection* in Q is a hyperbolic isometry other than the identity which fixes each point of Q.

Let Q_0 be the imaginary axis. By Theorem 1.2.8(iii), we have

$$\sinh^2[\tfrac{1}{2}\rho(z,w)] = \frac{|z-w|^2}{4\mathrm{Im}(z)\mathrm{Im}(w)}.$$

This shows that the map R_0: $z \to -\bar{z}$, a Euclidean reflection in Q_0, is a hyperbolic isometry; and since it fixes each point of Q_0, it is a hyperbolic reflection. If Q is another geodesic, there exists $T \in PSL(2,\mathbb{R})$ such that $T(Q)=Q_0$ (by Exercise 1.1). As T is a hyperbolic isometry, $T^{-1} \circ R_0 \circ T$ is the hyperbolic reflection fixing Q. As R_0 has order 2, every hyperbolic reflection has order 2.

Let m_i be a positive integer or ∞ ($i=1, 2, 3$) such that $\frac{1}{m_1}+\frac{1}{m_2}+\frac{1}{m_3}<1$, and τ be a hyperbolic triangle with vertices v_1, v_2, v_3, angles π/m_1, π/m_2, π/m_3 at these vertices, and sides M_1, M_2, M_3 opposite these vertices, as illustrated in Figure 19. Such a triangle exists by Exercise 4.7.

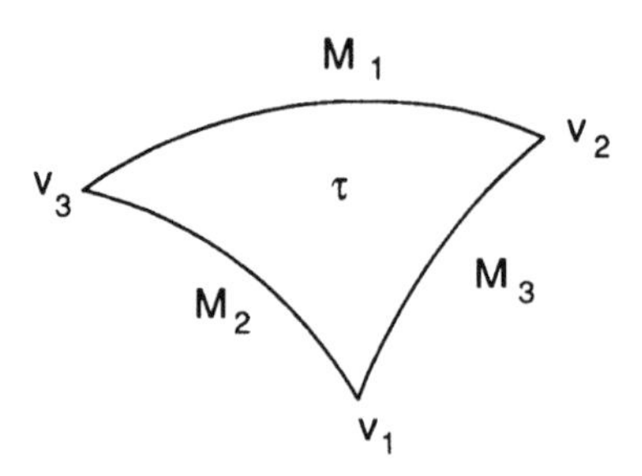

Fig. 19

Let R_i be the hyperbolic reflection in the geodesic containing M_i ($i=1,2,3$), and let Γ^* be the group generated by the reflections R_1, R_2, R_3. Since $R_i \notin PSL(2,\mathbb{R})$, Γ^* is not a Fuchsian group. However, consider $\Gamma=\Gamma^*\bigcap PSL(2,\mathbb{R})$. $\Gamma^*=\Gamma\bigcup\Gamma R_1$, for if $S\in\Gamma^*-\Gamma$, then SR_1 is the composition of two orientation-reversing isometries, so it is orientation-preserving and thus $SR_1\in PSL(2,\mathbb{R})$. Also, $SR_1\in\Gamma^*$ so that $SR_1\in\Gamma$, and $S=(SR_1)\ R_1\in\Gamma R_1$. It follows from Exercise 4.13 that $\left\{T(\tau)\mid T\in\Gamma^*\right\}$ form a *tessellation* of $\mathcal{H}$, that is every point of $\mathcal{H}$ belongs to some Γ^*-image of τ, and any two images of τ may only overlap by their boundary. It will follow that τ is a fundamental region for Γ^*. Now let p be any point in τ. The Γ^*-images of p are corresponding points of the other triangles of the tesselation, hence they form a discrete set. Thus the Γ-orbit of each point of $\mathcal{H}$ is a discrete set, and by Corollary 2.2.7, Γ is a Fuchsian group. It follows from Theorem 3.1.2 that $\tau\cup R_1(\tau)$ is a fundamental region for Γ (see Fig. 20). The sides $v_2v_1$and v_2v_1' are paired by R_1R_3 and the sides $v_3v_1$and v_3v_1' are paired by R_1R_2. Recalling that the angles of the original triangle τ are π/m_1, π/m_2, and π/m_3 and using Theorem 3.5.3, we see that $\{v_1,v_1'\}$ is an elliptic cycle both vertices of which are stabilized by cyclic groups of order m_1; $\{v_2\}$ is an elliptic cycle, v_2 being stabilized by a cyclic group of order m_2; and $\{v_3\}$ is an elliptic cycle, v_3 being stabilized by a cyclic group of order m_3.

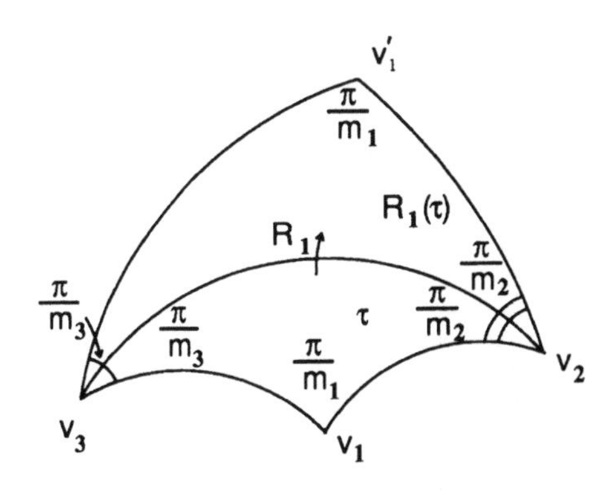

Fig. 20

Therefore $r=3$, $s=0$, and $\Gamma\backslash\mathcal{H}$ decomposes into 3 vertices, 2 edges (we have 2 pairs of sides in the fundamental region), and 1 simply connected face. By the Euler formula we have

$$2-2g=3-2+1=2,$$

giving $g=0$. (Alternatively, by "glueing" v_2v_1 to v_2v_1' and v_3v_1 to v_3v_1' we see that we obtain a surface homeomorphic to the sphere). Thus the signature of Γ is $(0; m_1, m_2, m_3)$.

4.5. Fuchsian groups of the first kind

Recall that a Fuchsian group Γ is of the first kind if its limit set $\Lambda(\Gamma)$ coincides with the set of points at infinity (see §3.4).

THEOREM 4.5.1. *If Γ is a geometrically finite Fuchsian group of the first kind, then Γ has a fundamental region of finite hyperbolic area.*

PROOF: Let F be a fundamental region for Γ with finitely many sides. If F has a free side, then any point of this side is not a limit point for Γ; hence Γ is not of the first kind. Thus F has no free sides. Then either F is cocompact, or it has a finite number of vertices at

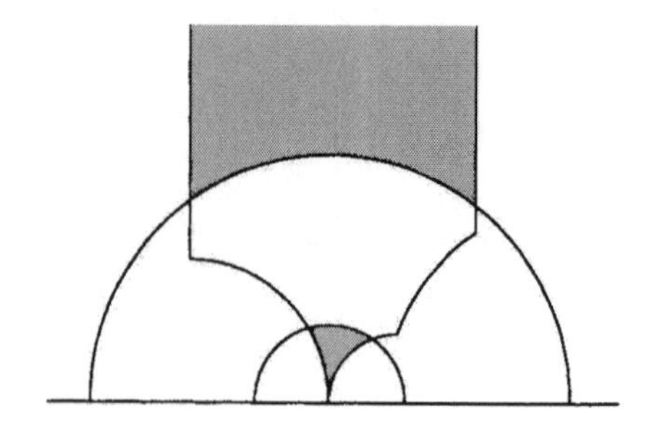

Fig. 21

infinity. Let as draw non-intersecting geodesics which separate the vertexes at infinity (see Fig. 21). Thus we divide F into a union of a compact polygon, and a finite number of hyperbolic triangles with one angle at infinity equal to 0 (shaded in Fig. 21). Since the hyperbolic area of each triangle is finite, the theorem follows. □

A converse statement is also true.

THEOREM 4.5.2. *If a Fuchsian group* Γ *has a fundamental region of finite hyperbolic area then* Γ *is of the first kind.*

PROOF: We shall use the unit disc model $\mathcal{U}$ to prove that each point of the principal circle Σ is a limit point for Γ. Let F be a Dirichlet fundamental region for Γ. Then $\mu(F)<\infty$, and by Theorem 3.5.1 F is locally finite.

LEMMA 4.5.3. *Let* $\Gamma=\{Id, g_1, g_2, ...\}$ *be a Fuchsian group acting on* $\mathcal{U}$, *and* F *be a locally finite fundamental region for* Γ. *Then Euclidean diameter of* $(g_n(F))\to 0$ *as* $n\to\infty$.

PROOF: If this were not so, we could find a subsequence $\{n_k\}$ and points z_k, $w_k \in g_{n_k}(F)$ such that $z_k\to z$, $w_k\to w$, $z\neq w$. Neither z nor w lie inside $\mathcal{U}$ since this would contradict the local finiteness of F, therefore $|z|=|w|=1$. Then $g_{n_k}(F)$ accumulate on the geodesic $[z,w]$, and this also contradicts the local finiteness. □

Let $z_0\in\Sigma$ and $U_\epsilon(z_0)$ be the intersection of the Euclidean disc of radius ϵ centered at z_0 with $\mathcal{U}$. Using Lemma 4.5.3, we can choose $A_1\in\Gamma$ such that $A_1(F)\subset U_\epsilon(z_0)$, since the hyperbolic area of $U_\epsilon(z_0)$ is infinite and F is a fundamental region. Letting $\epsilon\to 0$, we can choose a sequence of distinct transformations $\{T_n\}$ such that $T_n(z) \to z_0$ for any $z \in F$. □

REMARK: Another converse statement is Siegel's Theorem (Theorem 4.1.1).

4.6. Finitely generated Fuchsian groups

The goal of this section is to prove that finitely generated Fuchsian groups are geometrically finite. The converse to this statement is Theorem 3.5.4: it states that transformations which pair sides of a fundamental region F for a Fuchsian group Γ generate Γ. Hence if the number of sides of F is finite, then Γ is finitely generated.

THEOREM 4.6.1. *If Γ is finitely generated, it is geometrically finite.*

PROOF [B]: We shall use the unit disc model $\mathcal{U}$. Suppose that 0 is not a fixed point for any element in Γ, and let $F = D_0(\Gamma)$ be the Dirichlet region with center at the origin 0. We know (Theorem 3.5.4) that the transformations pairing the sides of F generate Γ. Each of finitely many generators of Γ is therefore a finite word in the side-pairing transformations of F. It follows that a finite number of side-pairing elements generate Γ: let these be $g_1, \dots, g_t$.

Choose some r in (0,1) such that the disc $\{|z| \leq r\}$ contains arcs of positive length of each of the sides paired by $g_1, \dots, g_t$. We always can choose r so that the circle $\{|z| = r\}$ does not meet any vertex of F and so that it is not tangent to any side of F. Let

$$K = F \cap \{|z| < r\}. \tag{4.6.1}$$

The two main steps in the proof are to show that:

(1) F can be expressed in the form

$$F=K\cup F_1\cup F_2\cup\ldots\cup F_n, \tag{4.6.2}$$

where K is the set defined in (4.6.1), which has compact closure, and each $\partial_0 F_i$ is connected.

(2) Only finitely many sides of F meet each F_i.

Since vertices of F are isolated in $\mathcal{U}$ (Exercise 3.10), we know that only finitely many sides of F meet K, and this will conclude the proof.

To derive the expression (4.6.2) we notice that

$$F\bigcap\{|z|=r\}=\sigma_1\cup\ldots\cup\sigma_s,$$

where σ_j are pairwise disjoint closed arcs of $\{|z|=r\}$ lying in F, and their end points lie on ∂F. Let

$$\Gamma(K)=\bigcup_{g\in\Gamma} g(K).$$

Observe that by construction of the region K, for each j, the set $K\bigcup g_j(K)$ is connected (K is convex) and hence so is each

$$K\bigcup g_{j_1}(K)\bigcup g_{j_1}g_{j_2}(K)\bigcup \ldots \bigcup g_{j_1}\cdots g_{j_q}(K)\ (1\leq j_i\leq t).$$

Since g_j generate Γ, this implies that $\Gamma(K)$ is connected.

Now recall that $F=D_0(\Gamma)$ coincides with the region R_0 for Γ (Theorem 3.3.5). If two sides of F are paired by an element $g_i\in\Gamma$ $(1\leq i\leq t)$ then they belong to isometric circles $I(g_i)$ and $I(g_i^{-1})$, and by Theorem 3.2.4, if $z\in I(g_i)$ and $w=g_i(z)$ then $|w|=|z|$. It follows that the collection of endpoints of all the σ_j are also paired by the side-pairing transformations. This implies that each endpoint of each σ_j is the endpoint of some $h(\sigma_i)$ for a unique h and unique σ_i, and we deduce that each σ_j lies in a simple arc G_j composed of images of the σ_i. Because there only finitely many σ_i, the arcs G_j contain images of the same σ_i; and the uniqueness of the construction of the G_j implies that G_j is invariant under some non-trivial element h_j of Γ. Note that G_j consists of the images of a compact arc under iterates of h_j.

Observe that a point of K cannot be congruent to any point of any σ_j since they belong to the same fundamental region, so $\Gamma(K)$ does not meet any G_j.

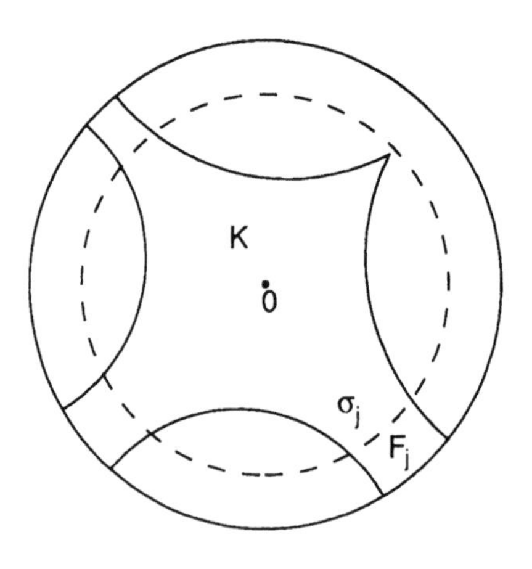

Fig. 22

Now let F_j be the union of σ_j and the component of $F-\sigma_j$ that does not contain the origin (see Fig. 22). Notice that G_j separates F_j and $\Gamma(K)$ in $\mathcal{U}$.

It is easy to see that $\partial_0 F_j$ is connected. Suppose u and v are two distinct point in this set. We have ru, rv$\in\sigma_j$. Since F is convex we can construct a curve τ_j by connecting u to ru radially, then ru to rv in σ_j, and finally v to rv radially. This curve lies in F_j and hence does not meet $\Gamma(K)$. If h ($\neq$Id) is in Γ, then h(F) does not meet τ_j (since τ_j lies inside $F_j \subset F$) and so lies on the same side of τ_j as does $\Gamma(K)$. Therefore the region Σ_j illustrated in Figure 23 does not meet any h(F), h$\neq$Id, and so lies in F. This shows that $\partial_0 F_j$ is connected, and (1) is established.

We now return to the classification of h_j stabilizing G_j and complete the proof. If h_j is elliptic, then G_j is a Jordan curve in $\mathcal{U}$ as illustrated in Figure 24, so one component of $\mathcal{U}-G_j$ has compact closure in $\mathcal{U}$. If this component contains F_j, then only finitely many sides of F meet F_j. If this component does not contain F_j, it contains $\Gamma(K)$ and so Γ is finite: then the Dirichlet polygon for Γ has only a finite number of sides.

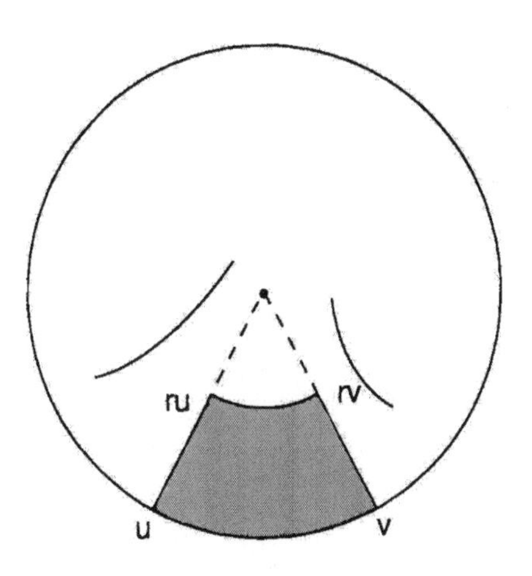

Fig. 23

If h_j is hyperbolic, then G_j is a cross-cut of $\mathcal{U}$ with the fixed points of h_j as its fixed points (see Fig. 25). One component of $\mathcal{U}-G_j$

contains $\Gamma(K)$ (and hence the orbit of the origin) and so every limit point of $\Gamma(0)$ and hence (by Lemma 4.5.3), every limit point of Γ lies in the Euclidean closure of this component. The other component of $\mathcal{U}-G_j$ contains F_j and there are no limit points on the open arc of Σ that bounds this component. However, F_j lies in F and hence lies between the perpendicular bisector of the segment $[0, h_j(0)]$ and that of the segment $[0, h_j^{-1}(0)]$. By Exercise 3.7, those perpendicular bisectors are the isometric circles $I(h_j^{-1})$ and $I(h_j)$ respectively. We observe that the fixed points of h_j lie inside the isometric circles $I(h_j)$ and $I(h_j^{-1})$ (since the points outside of $I(h_j)$ are mapped inside $I(h_j^{-1})$, and by Exercise 3.5(i) these circles do not intersect). We deduce that these isometric circles separate F_j from the fixed points of h_j, and that $\partial_0 F_j$ lies in the set of ordinary points of Γ. As the Euclidean diameters of images of F tend to zero (see Lemma 4.5.3 again) we see that F_j can meet only a finite number of images of F and hence only a finite number of sides of F.

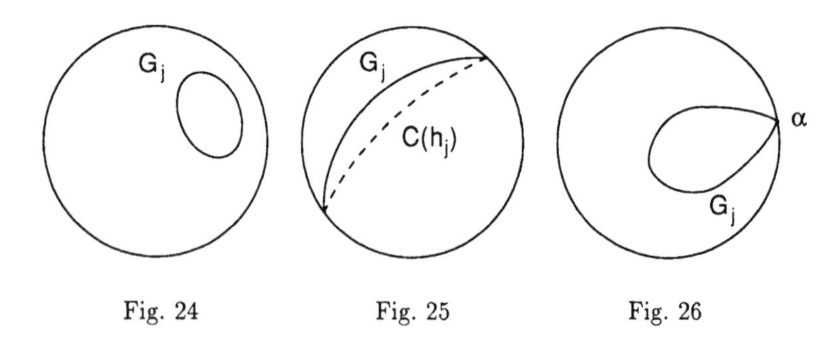

Fig. 24 Fig. 25 Fig. 26

Finally, suppose that h_j is parabolic. Then G_j is a closed Jordan curve in $\mathcal{U}$ (apart from its initial=final point α which is the fixed point of h_j and which lies on Σ as shown in Fig. 26). In this case one component of $\mathfrak{U}-G_j$ has the Euclidean closure consisting of the single point α. If this component contains $\Gamma(K)$, then repeating the argument for hyperbolic h_j, we conclude that $\Lambda(\Gamma)$ consists of a single point α. This implies that Γ is elementary, hence a cyclic parabolic subgroup whose fundamental region has two sides. If this component contains F_j, then either $\partial_0 F_j = \emptyset$, in which case only finitely many sides of F meet F_j, or $\partial_0 F_j$ consists of a single point α which is the fixed point of the parabolic element h_j. Recall that $F=R_0$, and that the isometric circles $I(h_j)$ and $I(h_j^{-1})$ are tangent to each other at the point α (Exercise 3.5(iii)). Therefore, in this case, the two sides of F meet at α, and again only finitely many sides of F meet F_j. □

EXERCISES FOR CHAPTER 4

4.1. Give another proof of Theorem 4.2.2(ii) using Theorem 4.1.1.

4.2. Show that the limit of $B_t(p)$ as $t \to \infty$ is a Euclidean circle passing through p and the end of the geodesic z(t) corresponding to $t=\infty$, and orthogonal to the geodesic z(t).

4.3. Let T be a non-parabolic transformation fixing a point b at infinity, $\omega(b)$ be a horocycle, $p \in \omega(b)$. Prove that either T(p) or $T^{-1}(p)$ lies inside $\omega(b)$.

4.4. Let Γ be a non-elementary Fuchsian group, F a locally finite fundamental region for Γ, and ξ a fixed point of some parabolic element in Γ, then there exists $T \in \Gamma$ s.t. $T(\xi) \in \partial_0(F)$.

4.5. Give a careful proof of Corollary 4.2.6.

4.6. Prove the Gauss-Bonnet formula for an n-sided star-like hyperbolic polygon Π with angles $\alpha_1, \dots, \alpha_n$:

$$\mu(\Pi) = (n-2)\pi - \sum_{i=1}^{n} \alpha_i.$$

4.7. Prove that for any non-negative real numbers α, β, γ such that $\alpha+\beta+\gamma<\pi$, there exists a hyperbolic triangle with angles α, β, γ.

4.8. Prove that for any $0 \le \alpha < \pi$ there exists an isoseles triangle with a given base and the angle between the equal sides equal to α.

4.9. If F is a compact Dirichlet region for a Fuchsian group then $\mu(F) \ge \pi/21$, the minimum being attained for the group with signature (0; 2, 3, 7).

4.10. If F is a noncompact Dirichlet region for a Fuchsian group with $\mu(F)<\infty$ then $\mu(F) \ge \pi/3$, the minimum being attained for the modular group which has signature (0; 2, 3; 1).

4.11. Show that for a Fuchsian group Γ with signature $(g; m_1, \dots, m_r; s)$,

$$\mu(\Gamma \backslash \mathcal{H}) = 2\pi\left[(2g-2) + \sum_{i=1}^{r}\left(1 - \frac{1}{m_i}\right) + s\right].$$

4.12. Prove that a hyperbolic reflection in Q is a restriction of a Euclidean inversion in Q to the upper half-plane if Q is a semicircle, or a Euclidean reflection if Q is a vertical line.

4.13. Show that $\{T(\tau) \mid T \in \Gamma^*\}$ forms a tessellation of $\mathcal{H}$.

5. ARITHMETIC FUCHSIAN GROUPS

5.1. Definitions of arithmetic Fuchsian groups

In this chapter we give some examples of Fuchsian groups. The most interesting and important ones are the so-called arithmetic Fuchsian groups, i.e. discrete subgroups of PSL(2,ℝ) obtained by some "arithmetic" constructions. One such construction we have already seen: if we choose all matrices of SL(2,ℝ) with *integer* coefficients, then the corresponding elements of PSL(2,ℝ) form the *modular group* PSL(2,ℤ). It is considered in detail in §§3.1, 3.2, 3.6, and 5.5 (see Example A). In §5.5 we study some important classes of its subgroups of finite index. The same construction, restriction of scalars to integers, allows us to obtain arithmetic subgroups of larger matrix groups, e.g. SL(n,ℤ) in SL(n,ℝ), Sp(2n,ℤ) in Sp(2n,ℝ), etc. The following natural definition of an arithmetic Fuchsian group that uses the above construction is given in [GP]. Let $g \to T(g)$ be a finite dimensional representation of the group PSL(2,ℝ). The elements of PSL(2,ℝ) which correspond to matrices T(g) with integer coefficients form a discrete subgroup of PSL(2,ℝ). All subgroups of PSL(2,ℝ) thus obtained and also their subgroups of finite index are called *arithmetic Fuchsian groups.* This definition is slightly different from a commonly used one of an arithmetic subgroup of a semisimple Lie group. The latter definition is given in the context of linear algebraic groups [BH], and is beyond the scope of this book. It is rather hard to check, except for trivial cases, whether or not a given Fuchsian group is arithmetic according to the above definition. However, it follows from results of A. Weil on classification of classical groups [W] (see also [Ti]) that the list of all arithmetic subgroups of SL(2,ℝ) is exhausted up to *commensurability* by Fuchsian groups *derived from quaternion algebras* over totally real number fields (for all definitions see § 5.2). If this field is $\mathbb{Q}$ and the quaternion algebra is isomorphic to $M(2,\mathbb{Q})$, the full matrix algebra over $\mathbb{Q}$, then the quotient space $\Gamma\backslash\mathcal{H}$ is not compact but has finite volume, and Γ is commensurable with the modular group (Theorem 5.5.12); in all other cases $\Gamma\backslash\mathcal{H}$ is compact (Theorems 5.2.6 and 5.4.1).

5.2. Fuchsian groups derived from quaternion algebras

Let F be a field of characteristic $\neq 2$. A *quaternion algebra* over F is an algebra A over F satisfying the following conditions:

(i) its *radical* R (i.e. an ideal $R \subset A$ such that $R^e=\{0\}$ for some integer e) is trivial,

(ii) its *center* $Z=\{x \in A \mid xy=yx \text{ for all } y \in A\}=F$,

(iii) $\dim_F(A)=4$.

An algebra satisfying (i)-(ii) is called *simple central algebra.*

Each quaternion algebra is isomorphic to an algebra $A=\left(\frac{a,b}{F}\right)$ with $a, b \in F^*=F-\{0\}$ and the basis $\{1, i, j, k\}$ where

$$i^2=a,\ j^2=b,\ k=ij=-ji. \tag{5.2.1}$$

In particular, a usual *Hamiltonian quaternion algebra* $\mathbb{H}$ is isomorphic to $\left(\frac{-1,-1}{\mathbb{R}}\right)$, and the matrix algebra $M(2,F)$ is isomorphic to $\left(\frac{1,1}{F}\right)$. Any element of the algebra A has the form

$$x=x_0+x_1i+x_2j+x_3k, \tag{5.2.2}$$

where $x_0, x_1, x_2, x_3 \in F$.

We see that quaternion algebras are *non-commutative.* If each element of A has an inverse, then a quaternion algebra is called a *division algebra.* We know that the matrix algebra $M(2,F)$ is not a division algebra since it has zero divisors. Theorem 5.2.4 below shows that the converse statement is also true. Division algebras are sometimes called *skew–fields.*

Let us define a linear map φ: A $\to$ M(2,F($\sqrt{a}$)) by sending the elements of the basis of A to the following matrices:

$$1 \to \begin{bmatrix} 1 & 0 \\ 0 & 1 \end{bmatrix}, \; i \to \begin{bmatrix} \sqrt{a} & 0 \\ 0 & -\sqrt{a} \end{bmatrix},$$
$$j \to \begin{bmatrix} 0 & 1 \\ b & 0 \end{bmatrix}, \; k \to \begin{bmatrix} 0 & \sqrt{a} \\ -b\sqrt{a} & 0 \end{bmatrix}. \tag{5.2.3}$$

Thus we have

$$\varphi(x)=g_x=\begin{bmatrix} x_0+x_1\sqrt{a} & x_2+x_3\sqrt{a} \\ b(x_2-x_3\sqrt{a}) & x_0-x_1\sqrt{a} \end{bmatrix}. \tag{5.2.4}$$

It is easy to check that the matrices in (5.2.3) are linearly independent, and that $\varphi(i^2)=a$, $\varphi(j^2)=b$, $\varphi(i)\varphi(j)=-\varphi(j)\varphi(i)$. We use Exercise 5.1 to conclude that φ is an isomorphism of the algebra A to an F–subalgebra of the algebra M(2, F($\sqrt{a}$)). If $a=t^2$ for some $t\in F^*$, the isomorphism φ is defined by the formula

$$\varphi(x)=g_x=\begin{bmatrix} x_0+x_1 t & x_2+x_3 t \\ b(x_2-x_3 t) & x_0-x_1 t \end{bmatrix}, \tag{5.2.5}$$

and it is onto M(2, F). Also notice that $F(\sqrt{a})\approx F(\sqrt{\lambda^2 a})$ for any $\lambda\in F^*$. Thus we have the following result:

THEOREM 5.2.1.
(i) If $a\in(F^*)^2$ *then* $A=\left(\frac{a,\, b}{F}\right)\approx M(2,F)$.
(ii) For any $\lambda\in F^*$, $\left(\frac{a,\, b}{F}\right)\approx\left(\frac{\lambda^2 a,\, b}{F}\right)$. □

DEFINITION. A field F is called *algebraically closed* if any polynomial in F[x] has all its roots in F.

COROLLARY 5.2.2. *If* F *is an algebraically closed field, then for any* a *and* b, $A=\left(\frac{a,b}{F}\right)\approx M(2,F)$. □

Let ι: A → A be the *standard involution* of A given by the formula $\iota(x)=\bar{x}$, where $\bar{x}=x_0-x_1i-x_2j-x_3k$ is called the conjugate of x. Then $\mathrm{Trd}(x)=x+\bar{x}=2x_0$ is called the *reduced trace* of x, and $\mathrm{Nrd}(x)=x\cdot\bar{x}=x_0{}^2-x_1{}^2a-x_2{}^2b+x_3{}^2ab$ is called the *reduced norm* of x. Notice that $\mathrm{Trd}(x)=\mathrm{Tr}(g_x)$ and $\mathrm{Nrd}(x)=\det(g_x)$, where $g_x=\varphi(x)$ is given by the formula (5.2.4). Obviously,

$$\mathrm{Nrd}(xy)=\mathrm{Nrd}(x)\mathrm{Nrd}(y), \quad \mathrm{Nrd}(1)=1.$$

THEOREM 5.2.3. A *is a division algebra if and only if* $\mathrm{Nrd}(x)=0$ *only for* $x=0$.

PROOF: Let $x\neq 0$. Then $\mathrm{Nrd}(x)\neq 0$. Since $\mathrm{Nrd}(x)=x\cdot\bar{x}$ we have $\frac{\bar{x}}{\mathrm{Nrd}(x)}\cdot x=1$ (i.e. $\frac{\bar{x}}{\mathrm{Nrd}(x)}$ is the element inverse to x), and A is a division algebra. Conversely, If A is a division algebra and $x\neq 0$, then $x^{-1}\neq 0$, and $\mathrm{Nrd}(x)\mathrm{Nrd}(x^{-1})=1$, i.e. $\mathrm{Nrd}(x)\neq 0$. □

THEOREM 5.2.4. *If* $A=\left(\frac{a,b}{F}\right)$ *is not isomorphic to* M(2,F) *then* A *is a division algebra.*

PROOF: First we notice that $a\notin(F^*)^2$, otherwise we would have $A\approx M(2,F)$ by Theorem 5.2.1. Then L=F(i) is a quadratic extension of F, and A=L+Lj. Suppose A is not a division algebra. Then by Theorem 5.2.3, there exists $h\in A$, $h\neq 0$ with $\mathrm{Nrd}(h)=0$. Let

$h=x_0+x_1i+x_2j+x_3k$. We have

$$0=\mathrm{Nrd}(h)=x_0^2-x_1^2a-x_2^2b+x_3^2ab=n(x_0+ix_1)-bn(x_2+ix_3), \quad (5.2.6)$$

where $n(z_0+iz_1)=z_0^2-az_1^2$ is the norm in the field L. We show first that $x_2+ix_3\neq 0$: for if $x_2+ix_3=0$ we have $n(x_0+ix_1)=0$, and since there are no zero divisors in the field L, $x_0+ix_1=0$; and then $h=0$, a contradiction. Thus we deduce from (5.2.6) that

$$b=\frac{n(x_0+ix_1)}{n(x_2+ix_3)}=n(q_0+iq_1),$$

where $q_0, q_1\in F$, i.e.

$$b=q_0^2-aq_1^2.$$

Now we construct a map from A onto M(2,F) sending the elements of the basis of A to the following matrices:

$$1\rightarrow\begin{bmatrix}1 & 0\\ 0 & 1\end{bmatrix}, i\rightarrow\begin{bmatrix}0 & 1\\ a & 0\end{bmatrix},$$

$$j\rightarrow\begin{bmatrix}q_0 & -q_1\\ q_1a & -q_0\end{bmatrix}, k\rightarrow\begin{bmatrix}q_1a & -q_0\\ aq_0 & -aq_1\end{bmatrix}. \quad (5.2.7)$$

It is easy to check that $i^2\rightarrow a$, $j^2\rightarrow b$, $ij=-ji$, and that the matrices in (5.2.7) are linearly independent. Hence $A\approx M(2,F)$, a contradiction.□

EXAMPLE D. Let $A=\left(\frac{5,11}{\mathbf{Q}}\right)$. Since for $h=1+3i+j+k\neq 0$, $\mathrm{Nrd}(h)=1^2-5\cdot 3^2-11\cdot 1^2+55\cdot 1^2=0$, A is not a division algebra. Hence, by Theorem 5.2.4, $A\approx M(2,\mathbf{Q})$.

The following theorem gives a series of examples of division algebras over **Q**.

THEOREM 5.2.5. *Let* b *be a prime number, and* a *be any quadratic non-residue* (mod b), *i.e.* $x^2\equiv a$ (mod b) *has no solutions in integers. Then the algebra* $A=\left(\frac{a,b}{\mathbf{Q}}\right)$ *is a division algebra.*

PROOF: Suppose not. Then by Theorem 5.2.3 there exists $x \in A$, $x \neq 0$ with norm

$$x_0^2 - x_1^2 a - x_2^2 b + x_3^2 ab = 0. \tag{5.2.8}$$

We may assume that x_0, x_1, x_2, x_3 have no common factors. It follows from (5.2.8) that

$$x_0^2 \equiv a x_1^2 \pmod{b}. \tag{5.2.9}$$

If b does not divide x_1, then x_1^2 is a quadratic residue mod b, and a product of a quadratic residue and a quadratic non-residue is a quadratic non-residue, a contradiction with (5.2.9). Thus $b|x_1$, and hence $b|x_0$, and we see from (5.2.8) that $x_2^2 \equiv a x_3^2 \pmod{b}$. By the same argument we conclude that $b|x_2$ and $b|x_3$ in contradiction with our assumption. □

If $A = \left(\frac{a,b}{F}\right)$ is a quaternion algebra over F, and σ: F → K is any homomorphism of F into another field K, we define $A^\sigma = \left(\frac{\sigma(a),\sigma(b)}{\sigma(F)}\right)$, and $A^\sigma \otimes K = \left(\frac{\sigma(a),\sigma(b)}{K}\right)$.

In what follows, F will be a *totally real algebraic number field of degree* n. This means that F is a field extension of $\mathbb{Q}$ of degree n, so that all n distinct embeddings of F into $\mathbb{C}$ are embeddings φ_i $(1 \leq i \leq n)$ into $\mathbb{R}$ where φ_1 is the identity. Let A be a quaternion algebra over F such that for $1 \leq i \leq n$ there exist $\mathbb{R}$-isomorphisms ρ_i,

$$\rho_1 \colon A^{\varphi_1} \otimes \mathbb{R} \to M(2, \mathbb{R}), \ \rho_i \colon A^{\varphi_i} \otimes \mathbb{R} \to \mathbb{H} \ \ (2 \leq i \leq n). \tag{5.2.10}$$

In this case one says that A *is unramified at the place* φ_1 and *ramified at all other infinite places* φ_i $(2 \leq i \leq n)$. Denote by $\mathrm{Nrd}_{\mathbb{H}}$ and $\mathrm{Trd}_{\mathbb{H}}$ the reduced norm and reduced trace in $\mathbb{H}$. Then for any $x \in A$ we have

$$\mathrm{Nrd}(x)=\det(\rho_1(x)),\ \mathrm{Trd}(x)=\mathrm{tr}(\rho_1(x)), \tag{5.2.11}$$

$$\varphi_i(\mathrm{Nrd}(x))=\mathrm{Nrd}_{\mathbb{H}}(\rho_i(x)),\ \varphi_i(\mathrm{Trd}(x))=\mathrm{Trd}_{\mathbb{H}}(\rho_i(x)). \tag{5.2.12}$$

EXAMPLE E. Let $\mathbb{H}=\left(\frac{-1,-1}{\mathbb{R}}\right)$ and $\mathbb{H}^1=\{x\in\mathbb{H} \mid \mathrm{Nrd}_{\mathbb{H}}(x)=1\}$ be the set of quaternions of reduced norm 1. Let us compute

$$\mathrm{Trd}_{\mathbb{H}}(\mathbb{H}^1)=\{\mathrm{Trd}_{\mathbb{H}}(x) \mid x\in\mathbb{H}^1\}.$$

We have

$$x=x_0+x_1i+x_2j+x_3k,$$

where $i^2=j^2=k^2=-1$, and $\mathrm{Nrd}(x)_{\mathbb{H}}=x_0{}^2+x_1{}^2+x_2{}^2+x_3{}^2=1$. Then $|x_0|\leq 1$, and hence $\mathrm{Trd}_{\mathbb{H}}(x)=2x_0\in[-2,\ 2]$. The converse statement is obviously also true, hence $\mathrm{Trd}_{\mathbb{H}}(\mathbb{H}^1)=[-2,\ 2]$.

THEOREM 5.2.6. *Let* A *be a quaternion algebra over a totally real number field* F *satisfying* (5.2.10). If $F\neq\mathbb{Q}$ *then* A *is a division algebra.*

PROOF: Suppose A is not a division algebra, then by Theorem 5.2.4 $A\approx M(2,\ F)\approx\left(\frac{1,1}{F}\right)$. Since $[F:\mathbb{Q}]=n\geq 2$, for any $i>1$ we have $A^{\varphi_i}\approx\left(\frac{1,1}{\varphi_i(F)}\right)\approx M(2,\varphi_i(F))$, and hence $A^{\varphi_i}\otimes\mathbb{R}\approx\left(\frac{1,1}{\mathbb{R}}\right)\approx M(2,\mathbb{R})$ not to $\mathbb{H}$, a contradiction with (5.2.10). □

Let $\mathcal{O}_F$ be the ring of integers in F. We give three equivalent definitions of an *order* in A.

DEFINITIONS. (I) An *order* $\mathcal{O}$ in A over F is a subring of A containing 1 which is a finitely generated $\mathcal{O}_F$-module generating the algebra A over F.

(II) An *order* $\mathcal{O}$ in A over F with $[F:\mathbb{Q}]=n$ is a subring of A which contains 1 and is a free $\mathbb{Z}$-module of rank 4n.

(III) An *order* $\mathcal{O}$ in A over F is a subring of *integer elements* in A (i.e. elements $x \in A$ such that $\mathrm{Nrd}(x) \in \mathcal{O}_F$ and $\mathrm{Trd}(x) \in \mathcal{O}_F$) such that $F \cdot \mathcal{O} = A$.

For example, it is obvious from Definition I that if $A=\left(\frac{a,b}{F}\right)$, and $a, b \in \mathcal{O}_F{}^*$, then

$$\mathcal{O}=\{x=x_0+x_1 i+x_2 j+x_3 k \mid x_1, x_2, x_3, x_4 \in \mathcal{O}_F\}$$

is an order in A.

We shall be interested in the *group of units in* $\mathcal{O}$ *of reduced norm* 1: $\mathcal{O}^1=\{x \in \mathcal{O} \mid \mathrm{Nrd}(x)=1\}$. For any order $\mathcal{O}$ in A, $\rho_1(\mathcal{O}^1)$ is a subgroup of $SL(2,\mathbb{R})$, and

$$\Gamma(A,\mathcal{O})=\rho_1(\mathcal{O}^1)/\{+1_2,-1_2\} \tag{5.2.13}$$

is a subgroup of $PSL(2,\mathbb{R})$.

THEOREM 5.2.7. *$\Gamma(A,\mathcal{O})$ is a Fuchsian group.*

PROOF: We shall give a proof in the case when A is a division algebra over $\mathbb{Q}$: $A=\left(\frac{a,b}{\mathbb{Q}}\right)$, $a>0$, $\mathcal{O}=\{x \in A \mid x_0, x_1, x_2, x_3 \in \mathbb{Z}\}$. Slightly abusing notations, we shall write $\mathcal{O}^1$ instead of $\rho_1(\mathcal{O}^1)$, and show that $\mathcal{O}_1$ is a discrete subgroup of $SL(2,\mathbb{R})$. The embedding of A into $SL(2,\mathbb{R})$ is defined in (5.2.4). It is sufficient to find a neighborhood of Id in $SL(2,\mathbb{R})$ which contains no elements of $\mathcal{O}^1$ different from Id. Let

$$U=\left\{g=\begin{bmatrix} g_{11} & g_{12} \\ g_{21} & g_{22} \end{bmatrix} \in SL(2,\mathbb{R}) \mid |g_{11}-1|<\tfrac{1}{2},\ |g_{12}|<\tfrac{1}{2},\right.$$
$$\left.|g_{21}|<\tfrac{1}{2},\ |g_{22}-1|<\tfrac{1}{2}\right\}. \tag{5.2.14}$$

Suppose $g_x \in \mathcal{O}^1 \cap U$. We have $g_{11}=x_0+x_1\sqrt{a}$, $g_{12}=x_2+x_3\sqrt{a}$, $g_{21}=b(x_2-x_3\sqrt{a})$, $g_{22}=x_0-x_1\sqrt{a}$ with $x_0, x_1, x_2, x_3 \in \mathbb{Z}$. It follows from (5.2.14) that $|g_{11}+g_{22}-2|<1$, i.e. $|2x_0-2|<1$, which implies $x_0=1$. Also since $b>1$, we have $|x_2-x_3\sqrt{a}|<\frac{1}{2b}<\frac{1}{2}$, and thus $|2x_2|<1$, which implies $x_2=0$. We also have $|x_1\sqrt{a}|<\frac{1}{2}$, $|x_3\sqrt{a}|<\frac{1}{2}$, which implies $x_1=x_3=0$, and thus $g_x=\mathrm{Id}$. Thus $\Gamma(A, \mathcal{O})$ is a discrete subgroup of $PSL(2,\mathbb{R})$, or a Fuchsian group. □

DEFINITION. If Γ is a subgroup of finite index of some $\Gamma(A,\mathcal{O})$, then we call Γ a *Fuchsian group derived from a quaternion algebra* A.

DEFINITION. Two groups are called *commensurable* if their intersection has finite index in each of them.

DEFINITION. If Γ is commensurable with some $\Gamma(A,\mathcal{O})$, then Γ is called *an arithmetic Fuchsian group.*

5.3. Criteria for arithmeticity

Here we give a characterization of arithmetic Fuchsian groups Γ in terms of the set $\mathrm{tr}(\Gamma)=\{\pm \mathrm{Tr}(T) \mid T\in\Gamma\}$. This is due to K.Takeuchi [T1, T2]. For a field K, let

$$SL(2,F)=\left\{\begin{bmatrix} a & b \\ c & d \end{bmatrix} \mid a, b, c, d\in K,\ ad-bc=1\right\}$$

and

$$PSL(2,K)=SL(2,K)/\{\pm 1_2\}.$$

LEMMA 5.3.1. *Let* Γ *be a Fuchsian group such that* $\mu(\Gamma\backslash\mathcal{H})<\infty$ *and the set* $\mathrm{tr}(\Gamma)$ *is contained in an algebraic number field* k, $[k : \mathbb{Q}]<\infty$. *Then there exists* $g\in SL(2,\mathbb{R})$ *and an algebraic number field* K, $[K : \mathbb{Q}]<\infty$ *such that* $g^{-1}\Gamma g\subseteq PSL(2,K)$.

PROOF: By Theorem 4.5.2, Γ is a Fuchsian group of the first kind, hence non-elementary (see Exercise 3.8); hence it contains a hyperbolic transformation T. Let us denote by e_1, e_2 eigenvectors of a matrix representing T, and by λ, λ^{-1} their respective eigenvalues. Since $\mathrm{Tr}(T)>2$, λ is a real number (see §2.1). We can choose e_1 and e_2 such that $\det(e_1,e_2)>0$. Let $g_1=\frac{1}{\sqrt{\det(e_1,e_2)}}(e_1,e_2)$, and $K=k(\lambda)$. (Notice that although λ is only defined up to a sign, $K=k(\lambda)$ is well-defined.) Then $g_1^{-1}Tg_1=\begin{bmatrix} \lambda & 0 \\ 0 & \lambda^{-1} \end{bmatrix}$. Take an element $T_1\in\Gamma$ such that $g_1^{-1}T_1g_1=\begin{bmatrix} a & b \\ c & d \end{bmatrix}$ with $c\neq 0$, $b>0$ (this can be done because Γ is not elementary), and let $g_2=\begin{bmatrix} \sqrt{b} & 0 \\ 0 & 1/\sqrt{b} \end{bmatrix}$. Then we know that $(g_1g_2)^{-1}\Gamma(g_1g_2)$ contains two elements:

$$T_0=\begin{bmatrix} \lambda & 0 \\ 0 & \lambda^{-1} \end{bmatrix}(\lambda\neq 1) \text{ and } T_1=\begin{bmatrix} a_1 & 1 \\ c_1 & d_1 \end{bmatrix}(c_1\neq 0).$$

Take any element $T=\begin{bmatrix} a & b \\ c & d \end{bmatrix}$ of $(g_1g_2)^{-1}\Gamma(g_1g_2)$. Then

$$\begin{bmatrix} \lambda & 0 \\ 0 & \lambda^{-1} \end{bmatrix}\begin{bmatrix} a & b \\ c & d \end{bmatrix}=\begin{bmatrix} \lambda a & \lambda b \\ \lambda^{-1}c & \lambda^{-1}d \end{bmatrix} \tag{5.3.1}$$

is in $(g_1g_2)^{-1}\Gamma(g_1g_2)$, and hence $\lambda a+\lambda^{-1}d$ as well as $a+d$ are in K. Hence a and d are in K. In particular, a_1 and d_1 are in K. Since $\det(T_1)=1$, c_1 is also in K. We also have the following relation

$$\begin{bmatrix} a & b \\ c & d \end{bmatrix}\begin{bmatrix} a_1 & 1 \\ c_1 & d_1 \end{bmatrix}=\begin{bmatrix} aa_1+bc_1 & a+bd_1 \\ ca_1+dc_1 & c+dd_1 \end{bmatrix}, \tag{5.3.2}$$

which implies $aa_1+bc_1\in K$ and $c+dd_1\in K$, and since $a, a_1, d, d_1, c_1\in K$ we have $b, c\in K$. □

THEOREM 5.3.2. *Let the assumption be as in Lemma 5.3.1. Let* $k_0=\mathbb{Q}(\mathrm{tr}(T)\ |T\in\Gamma)$ *and* $A=k_0[\Gamma]=\{\sum_{i=1}^{d} a_i T_i \mid a_i\in k_0,\ T_i\in\Gamma\}$. *Then* A *is a quaternion algebra over* k_0.

PROOF: By Lemma 5.3.1, we may assume that Γ contains two elements

$$T_0=\begin{bmatrix} \lambda & 0 \\ 0 & \lambda^{-1} \end{bmatrix} (\lambda\neq 1) \text{ and } T_1=\begin{bmatrix} a_1 & 1 \\ c_1 & d_1 \end{bmatrix} (c_1\neq 0),$$

and $\Gamma\subseteq PSL(2,K_0)$, where $K_0=k_0(\lambda)$ is either k_0 or a quadratic extension of k_0. Hence $A\subseteq M(2, K_0)$, and $1<\dim_{k_0}(A)\leq 8$. In order to prove that A is a quaternion algebra we have to show that

(i) its radical R is trivial,

(ii) its center $Z=k_0$,

(iii) $\dim_{k_0}(A)=4$.

Let $T=\begin{bmatrix} a & b \\ c & d \end{bmatrix}\in R$, $T^e=0$. Then $\det(T)=0$ and $\mathrm{tr}(T^j)=\mathrm{tr}(T)^j$ for $1\leq j\leq e$, which implies $\mathrm{tr}(T)=0$. Since R is an ideal, by (5.3.1) $a+d=0$ and $\lambda a+\lambda^{-1}d=0$, hence $a=d=0$, and by (5.3.2) $b=c=0$ as well. This shows that $R=\{0\}$. Now let $T=\begin{bmatrix} a & b \\ c & d \end{bmatrix}\in Z$. Since T commutes with T_0, $c=b=0$ (by Theorem 2.3.3). Since T commutes with T_1, we have $a=d$, hence $T=a\cdot\mathrm{Id}$. But for any $T'\in A$, $\mathrm{tr}(T')\in k_0$, hence $a=\frac{1}{2}\mathrm{tr}(T)\in k_0$. Hence A is a central simple algebra. Its dimension is a square of an integer, hence it must be equal to 4. □

LEMMA 5.3.3. *Let* Γ *be a Fuchsian group with* $\mu(\Gamma\backslash\mathcal{H})<\infty$, $k_0=\mathbb{Q}(\mathrm{tr}(T)\ |T\in\Gamma)$, $[k_0:\mathbb{Q}]<\infty$, *and* $\mathrm{tr}(\Gamma)$ *is contained in the ring of integers of* k_0 *denoted by* $\mathcal{O}_{k_0}$. *Put*

$$A=k_0[\Gamma]=\{\sum_{i=1}^{d} a_i T_i \mid a_i\in k_0,\ T_i\in\Gamma\} \tag{5.3.3}$$

and $$\mathcal{O}=\mathcal{O}_{k_0}[\Gamma]=\{\sum_{i=1}^{d} a_i T_i \mid a_i\in\mathcal{O}_{k_0},\ T_i\in\Gamma\}. \tag{5.3.4}$$

Then $\mathcal{O}$ is an order of the quaternion algebra A.

PROOF: It is trivial that $\mathcal{O}$ is a subring containing the identity, and that it generates the algebra A over k_0. We only need to show that $\mathcal{O}$ is a finitely generated $\mathcal{O}_{k_0}$-module. By Lemma 5.3.1 we may assume that Γ contains two elements:

$$T_0=\begin{bmatrix} \lambda & 0 \\ 0 & \lambda^{-1} \end{bmatrix} (\lambda\neq 1) \text{ and } T_1=\begin{bmatrix} a_1 & 1 \\ c_1 & d_1 \end{bmatrix} (c_1\neq 0),$$

and that $\Gamma\subseteq PSL(2,K_0)$ where $K_0=k_0(\lambda)$. Take an arbitrary element $T=\begin{bmatrix} a & b \\ c & d \end{bmatrix}\in\mathcal{O}$. Then by (5.3.1), $a+d$ and $\lambda a+\lambda^{-1}d$ are in $\mathcal{O}_{k_0}$. Notice that λ and λ^{-1} are units in K_0, and $\mathcal{O}_{k_0}$ is a subring of the ring of integers of K_0. Hence a and d are in the fractional ideal $\frac{1}{\lambda^2-1}\mathcal{O}_{k_0}$ of K_0. By (5.3.2) aa_1+bc_1 and $c+dd_1$ are also in $\frac{1}{\lambda^2-1}\mathcal{O}_{k_0}$. Thus, for any $T\in\Gamma$, all its coefficients are contained in a fractional ideal $\frac{1}{\lambda^2-1}\mathcal{O}_{k_0}$. Hence $\mathcal{O}$ is a submodule of a free $\mathcal{O}_{k_0}$-module of rank 4, and therefore is a finitely generated $\mathcal{O}_{k_0}$-module. □

Now we can state necessary and sufficient conditions for arithmeticity of Fuchsian groups.

THEOREM 5.3.4. *Let Γ be a Fuchsian group with $\mu(\Gamma\backslash\mathcal{H})<\infty$. Then Γ is derived from a quaternion algebra* A *over a totally real number field* F if and only if Γ satisfies the following conditions:

(I) *Let k_1 be the field $\mathbb{Q}(tr(T)\ |T\in\Gamma)$ generated by the the set $tr(\Gamma)$ over $\mathbb{Q}$. Then k_1 is an algebraic number field of finite degree, and $tr(\Gamma)$ is contained in the ring of integers of k_1, $\mathcal{O}_{k_1}$.*

(II) Let φ be any embedding of k_1 into $\mathbb{C}$ such that $\varphi\neq$the identity.

Then $\varphi(\mathrm{tr}(\Gamma))$ *is bounded in* $\mathbb{C}$.

PROOF: Necessity of the conditions (I) and (II). Let Γ be a subgroup of finite index in $\Gamma(A, \mathcal{O})$ where $\mathcal{O}$ is an order in A. For any $T\in\Gamma$, $\mathrm{tr}(T)\in F$. Therefore k_1 is contained in F and k_1 is also totally real. Since $\mathrm{Trd}(\mathcal{O})$ is contained in $\mathcal{O}_F$, we see that $\mathrm{tr}(\Gamma)$ is contained in $\mathcal{O}_{k_1}$, and condition (I) is satisfied. Now suppose that $n\geq 2$. By (5.2.12) we see that for $2\leq i\leq n$, $\varphi_i(\mathrm{tr}(\Gamma))$ is contained in $\mathrm{Trd}_{\mathbb{H}}(\rho_i(\mathcal{O}^1))$. On the other hand, for any $x\in\mathcal{O}^1$, we have $\mathrm{Nrd}_{\mathbb{H}}(\rho_i(x))=\varphi_i(\mathrm{Nrd}(x))$. Hence $\rho_i(\mathcal{O}^1)$ is contained in the set $\mathbb{H}^1=\{x\in\mathbb{H} \mid \mathrm{Nrd}_{\mathbb{H}}(x)=1\}$. But by Example E of §5.2 we know that $\mathrm{Trd}_{\mathbb{H}}(\mathbb{H}^1)$ coincides with the interval $[-2, 2]$, so $\varphi_i(\mathrm{tr}(\Gamma))$ is bounded in $\mathbb{R}$ for $2\leq i\leq n$. It remains to show that $k_1=F$. Suppose F is a proper extension of k_1. Then for some i ($2\leq i\leq n$), $\varphi_i|_{k_1}=$the identity. Using this φ_i and the definition of k_1 we see that $\mathrm{tr}(\Gamma)=\varphi_i(\mathrm{tr}(\Gamma))$ is contained in the interval $[-2, 2]$. This means that Γ containes no hyperbolic elements, a contradiction with the fact that Γ is non-elementary (see Theorem 4.5.2 and Exercise 3.8).

Sufficiency of the conditions (I) and (II). First, using Lemma 5.3.3 we construct a quaternion algebra $A(\Gamma)$ over $k_1=\mathbb{Q}(\mathrm{tr}(T) \mid T\in\Gamma)$, and an order $\mathcal{O}(\Gamma)$ by formulae (5.3.3) and (5.3.4).

LEMMA 5.3.5. *Suppose* Γ *is a Fuchsian group with* $\mu(\Gamma\backslash\mathcal{H})<\infty$ *and satisfying (I) and (II) of Theorem* 5.3.4. *Then* $k_1=\mathbb{Q}(\mathrm{tr}(T) \mid T\in\Gamma)$ *is a totally real number field. Moreover, if* φ *is any embedding of* k_1 into $\mathbb{R}$ *such that* $\varphi\neq$*the identity, then* $\varphi(\mathrm{tr}(\Gamma))$ *is contained in the interval* $[-2, 2]$.

PROOF: Take any $T\in\Gamma$, and let u and 1/u be the eigenvalues of a matrix representing T. Let φ be any embedding of k_1 into $\mathbb{C}$, $\varphi\neq$the identity. Extend φ to an isomorphism ψ of $k_1(u)$ into $\mathbb{C}$. We shall show that $|\psi(u)|=1$. Suppose $|\psi(u)|\neq 1$. Then by the inequality

$$|\varphi(\mathrm{tr}(T^m))|=|(\psi(u))^m+1/(\psi(u))^m|\geq||\psi(u)|^m-1/|\psi(u)|^m|,$$

the set $\{\varphi(\mathrm{tr}(T^m)) \mid m\in\mathbb{Z}\}$ is not bounded, which contradicts (II). Hence $|\psi(u)|=1$. Then

$$\varphi(\mathrm{tr}(T))=\psi(u)+1/\psi(u)=\psi(u)+\overline{\psi(u)}.$$

Hence $\varphi(\mathrm{tr}(T))$ is a real number contained in the interval $[-2, 2]$. □

PROPOSITION 5.3.6. *Let* Γ *be a Fuchsian group with* $\mu(\Gamma\backslash\mathcal{H})<\infty$, *satistying (I) and (II) of Theorem 5.3.4. Then* $A(\Gamma)$ *satisfies* (5.2.10).

PROOF: In view of Lemma 5.3.1 we may assume that Γ contains the following two elements:

$$T_0=\begin{bmatrix} \lambda & 0 \\ 0 & \lambda^{-1} \end{bmatrix} (\lambda\neq 1) \text{ and } T_1=\begin{bmatrix} a_1 & 1 \\ c_1 & d_1 \end{bmatrix} (c_1\neq 0).$$

We shall show that $K=k_1(\lambda)$ is a proper quadratic extension of k_1. If k_1 is a proper extension of $\mathbb{Q}$, then there exists an embedding ψ: $K\to\mathbb{C}$ such that $\psi|_{k_1}\neq$the identity. $\psi(\lambda)$ and $1/\psi(\lambda)$ are the roots of the equation $x^2-\psi(t_0)x+1=0$, where $t_0=\mathrm{tr}(T_0)$. By Lemma 5.3.5 we have $|\psi(t_0)|<2$. Therefore $\psi(K)=\psi(k_1(\lambda))$ is an imaginary field. On the other hand, since k_1 is totally real, $\psi(k_1)$ is a real field; hence K does not coincide with k_1. If $k_1=\mathbb{Q}$, then t_0 is a rational integer such that $|t_0|>2$. Therefore the polynomial x^2-t_0x+1 is irreducible over $\mathbb{Q}$, and hence K is a proper extension of k_1. For $a\in K$, let a' be its k_1-conjugate. Then $1/\lambda=\lambda'$. We know that

$$\mathrm{tr}(T_1)=a_1+d_1\in k_1, \tag{5.3.5}$$

and

$$\mathrm{tr}(T_0T_1)=a_1\lambda+d_1\lambda'\in k_1. \qquad (5.3.6)$$

Since λ and λ' are linearly independent over k_1 we can write uniquely $a_1=\alpha_0\lambda+\alpha_1\lambda'$, $d_1=\delta_0\lambda+\delta_1\lambda'$. From (5.3.6) we have

$$(\alpha_0\lambda+\alpha_1\lambda')\lambda+(\delta_0\lambda+\delta_1\lambda')\lambda'=(\alpha_0\lambda'+\alpha_1\lambda)\lambda'+(\delta_0\lambda'+\delta_1\lambda)\lambda,$$

and hence $(\alpha_0-\delta_1)\lambda^2+(\delta_1-\alpha_0)(\lambda')^2=0$ which implies $\alpha_0=\delta_1$. From (5.3.5) we have $\alpha_0\lambda+\alpha_1\lambda'+\delta_0\lambda+\delta_1\lambda'=\alpha_0\lambda'+\alpha_1\lambda+\delta_0\lambda'+\delta_1\lambda$, hence $\alpha_0+\delta_0-\alpha_1-\delta_1=0$, and finally $\alpha_1=\delta_0$. This implies $d_1=a_1'$, and since $\det(T_1)=1$, we also obtain $c_1\in k_1$. Consequently we can rewrite

$$T_0=\begin{bmatrix}\lambda & 0\\ 0 & \lambda'\end{bmatrix}(\lambda\neq 1) \text{ and } T_1=\begin{bmatrix}a_1 & 1\\ c_1 & a_1'\end{bmatrix}(c_1\neq 0,\ c_1\in k_1).$$

We also notice that 1_2, T_0, T_1, T_0T_1 are linearly independent, and hence form a basis for $A(\Gamma)$ over k_1, and we see that

$$A(\Gamma)=\left\{\begin{bmatrix}a & b\\ b'c_1 & a'\end{bmatrix} \mid a, b\in K,\ c_1\in k_1\right\}.$$

LEMMA 5.3.7. *Let ψ be any embedding of $K=k_1(\lambda)$ into $\mathbb{C}$ such that $\psi|k_1\neq$ the identity. Then*

(i) for any element $T=\begin{bmatrix}a & b\\ b'c_1 & a'\end{bmatrix}$ of Γ we have $|\psi(a)|\leq 1$.

(ii) for $T_1=\begin{bmatrix}a_1 & 1\\ c_1 & a_1'\end{bmatrix}$ we have $\psi(c_1)<0$.

PROOF of (i): By Lemma 5.3.5, for any $T=\begin{bmatrix}a & b\\ b'c_1 & a'\end{bmatrix}\in\Gamma$ we have the inequality $|\psi(\mathrm{tr}(TT_0^m))|\leq 2$. Notice that for any $a=\alpha_0\lambda+\alpha_1\lambda'\in K$, $\psi(a')=\psi(\alpha_0)\psi(\lambda')+\psi(\alpha_1)\psi(\lambda)=\overline{\psi(\alpha_0)\psi(\lambda)+\psi(\alpha_1)\psi(\lambda')}=\overline{\psi(a)}$, since $|\psi(\lambda)|=1$ by the proof of Lemma 5.3.5 and k_1 is totally real. We have $\psi(\mathrm{tr}(TT_0^m))=\psi(a\lambda^m)+\psi(a'(\lambda')^m)=\psi(a\lambda^m)+\overline{\psi(a\lambda^m)}=2\mathrm{Re}(\psi(a)\cdot\psi(\lambda^m))$.

Since $\psi(\lambda)$ is not a root of unity, the set $\{\psi(\lambda)^m \mid m\in\mathbb{Z}\}$ is a dense subgroup of the unit circle S^1 (see Lemma 2.2.2(ii)). Therefore we have $|\mathrm{Re}(\psi(a)\cdot z)|\le 1$ for any $z\in S^1$ which implies $|\psi(a)|\le 1$.

PROOF of (ii): Applying (i) to T_1 we see that $|\psi(a_1)|\le 1$. By the equation $\det(T_1)=a_1a_1'-c_1=1$ we have $\psi(c_1)=\psi(a_1a_1')-1 =|\psi(a_1)|^2-1\le 0$. By the fact that $c_1\ne 0$, we see that $\psi(c_1)<0$. □

Now we can finish the proof of Proposition 5.3.6. Let $\{\varphi_i\}$ $(1\le i\le n)$ be all distinct embeddings of k_1 into $\mathbb{R}$, and assume that φ_1=the identity. Extend φ_i to an isomorphism ψ_i of $K=k_1(\lambda)$ into $\mathbb{C}$, and define an embedding Ψ_i of $A(\Gamma)$ into $M(2,\mathbb{R})$ in the following way:

$$\Psi_i : \alpha=\begin{bmatrix} a & b \\ b'c_1 & a' \end{bmatrix} \to \Psi_i(\alpha)=\begin{bmatrix} \psi_i(a) & \psi_i(b) \\ \psi_i(b'c_1) & \psi_i(a') \end{bmatrix}.$$

Then $A^{\varphi_i}=A^{\psi_i}=\Psi_i(A(\Gamma))$ is a quaternion algebra over $\psi_i(k_1)=\varphi_i(k_1)$. We have $A^{\varphi_1}\otimes\mathbb{R}\approx M(2,\mathbb{R})$. As we have seen in the proof of Lemma 5.3.7 $\psi_i(a')=\overline{\psi_i(a)}$ for $2\le i\le n$ we conclude that

$$A^{\varphi_i}=\left\{\begin{bmatrix} a & b \\ \bar{b}\psi_i(c_1) & \bar{a} \end{bmatrix} \mid a,\, b \in \psi_i(K),\ c_1\in k_1\right\}.$$

Using Lemma 5.3.7, one can prove (see Exercise 5.5) that $A^{\varphi_i}\otimes\mathbb{R}\approx\mathbb{H}$.□

To finish the proof of Theorem 5.3.4, we notice that (by Lemma 5.3.3, Lemma 5.3.5, and Proposition 5.3.6) k_1, $A(\Gamma)$, and $\mathcal{O}(\Gamma)$ satisfy the assumptions of §5.2. Clearly, Γ is a subgroup of $\Gamma(A(\Gamma),\,\mathcal{O}(\Gamma))$. Since both $\Gamma\backslash\mathcal{H}$ and $\Gamma(A(\Gamma),\,\mathcal{O}(\Gamma))\backslash\mathcal{H}$ are of finite volume, Γ is a subgroup of finite index in $\Gamma(A(\Gamma),\,\mathcal{O}(\Gamma))$. This shows that Γ is a Fuchsian group derived from a quaternion algebra. □

The following result follows immediately from Theorem 5.3.4: it characterizes Fuchsian groups derived from a quaternion algebras over $\mathbb{Q}$.

THEOREM 5.3.8. *Let* Γ *be a Fuchsian group with* $\mu(\Gamma\backslash\mathcal{H})<\infty$. *Then* Γ *is derived from a quaternion algebra over* $\mathbb{Q}$ *if and only if for every* $T\in\Gamma$, $tr(T)\in\mathbb{Z}$.

The following theorem is a very general fact about matrices with integral traces.

THEOREM 5.3.9. *Let* $T\in PSL(2,\mathbb{R})$ *be an elliptic element, and let* $tr(T)\in\mathbb{Z}$, *then* T may be only of order 2 *or* 3.

PROOF: We have $T(z)=\frac{az+b}{cz+d}$ $(ad-bc=1)$. Since T is elliptic, $tr(T)=|a+d|<2$. Therefore we have only three possibilities: $a+d=0$, 1, -1. The matrix $A=\begin{bmatrix} a & b \\ c & d \end{bmatrix}$ satisfies its characteristic equation: $x^2-(a+d)x+1=0$. If $a+d=0$ we have

$$A^2+1_2=0, \text{ i.e. } A^2=-1_2.$$

If $a+d=1$, we have

$$A^2-A+1_2=0, \text{ i.e. } A^2=A-1_2, \text{ and } A^3=A^2-A=-1_2.$$

If $a+d=-1$, we have

$$A^2+A+1_2=0, \text{ i.e. } A^2=-A-1_2, \text{ and } A^3=-A^2-A=1_2.$$

The theorem follows now from the fact that the identity in $PSL(2,\mathbb{R})$ is $\{1_2,-1_2\}$. □

COROLLARY 5.3.10. *A Fuchsian group derived from a quaternion algebra may have elliptic elements only of orders* 2 *and* 3.

Let Γ be a Fuchsian group with $\mu(\Gamma\backslash\mathcal{H})<\infty$. Then by Siegel's Theorem (Theorem 4.1.1), it is geometrically finite; hence (Theorem 3.5.4) is finitely generated. Denote by $\Gamma^{(2)}$ the subgroup of Γ generated by the set $\{T^2 \mid T\in\Gamma\}$. By Exercise 5.6, $\Gamma/\Gamma^{(2)}$ is a finite abelian group of type $(2, 2, \cdots, 2)$. Therefore $\Gamma^{(2)}$ is a subgroup of Γ of finite index. If $\Gamma^{(2)}$ is derived from a quaternion algebra, then being commensurable with $\Gamma^{(2)}$, Γ is an arithmetic Fuchsian group. The converse statement is also true [T1]:

THEOREM 5.3.11. *Γ is an arithmetic Fuchsian group if and only if $\Gamma^{(2)}$ is derived from a quaternion algebra.*

5.4. Compactness of $\Gamma\backslash\mathcal{H}$ for Fuchsian groups derived from division quaternion algebras

We know that for the modular group the quotient space has finite volume but is not compact (see §3.1). We shall see in §5.5 (Theorem 5.5.12) that the same is true for its subgroups of finite index, i.e. for Fuchsian groups derived from the matrix quaternion algebra $M(2,\mathbb{Q})$. The following theorem shows that for all other arithmetic Fuchsian groups Γ the factor space $\Gamma\backslash\mathcal{H}$ is compact [BH].

THEOREM 5.4.1. *Suppose a Fuchsian group Γ is derived from a division quaternion algebra. Then the quotient space $\Gamma\backslash\mathcal{H}$ is compact.*

PROOF: We shall give a proof [GP] in the simplest case when A is a division algebra over $\mathbb{Q}$: $A=\left(\frac{a,b}{\mathbb{Q}}\right)$, $a>0$, $\mathcal{O}=\{x\in A \mid x_0, x_1, x_2, x_3\in\mathbb{Z}\}$. We shall use the following general lemma about lattices in $\mathbb{R}^n$.

LEMMA 5.4.2. (Minkowski's lemma) *Let L be a lattice in $\mathbb{R}^n$, i.e. $L=\Big\{(x_1, \dots, x_n) \mid x_i=\sum_{j=1}^{n} x_{ij}n_j$, x_{ij} are fixed real numbers, n_j range* *over* $\mathbb{Z}\}$. *Suppose that the determinant* $|x_{ij}|=\Delta\neq 0$. *Then any convex symmetric with respect to* 0 *domain in* $\mathbb{R}^n$ *of volume* $\geq 2^n\Delta$ *contains at least two (symmetric with respect to* 0*) points of* L *besides* 0.

PROOF of LEMMA: Suppose U is such a domain which contains no other point of L besides 0. Let R: $\mathbb{R}^n \to \mathbb{R}^n$ be given by $R(x)=\frac{1}{2}x$. Let $U_0=R(U)$. We define for $S,A\in\mathbb{R}^n$, $p_{SA}:\mathbb{R}^n \to \mathbb{R}^n$ to be a translation of $\mathbb{R}^n$ along the vector $\overrightarrow{SA}$, and for all $A\in L$ let $U_A=p_{0A}(U_0)$. We shall show that $U_A\cap U_B=\emptyset$ for $A\neq B$. Suppose we have $C\in U_A\cap U_B$(see Fig 27). Let D be a point symmetric to C with respect to A. Since U_A is symmetric and $C\in U_A$, $D\in U_A$. Let $C'=p_{AD}(B)$. We have $U_B=p_{AB}(U_A)$, therefore $C'\in U_B$; and since U_B is convex, the middle point of CC′, say E, also belongs to U_B. Similarly, $E\in U_A$. Notice that C′ belongs to the plane through the points A, B, C. Hence ACBC′ is a parallelogram, and thus E is also the mid-point of AB. Thus $B\in R^{-1}(U_A)$ which is congruent to U, so U contains a point of L different from 0, a contradiction.

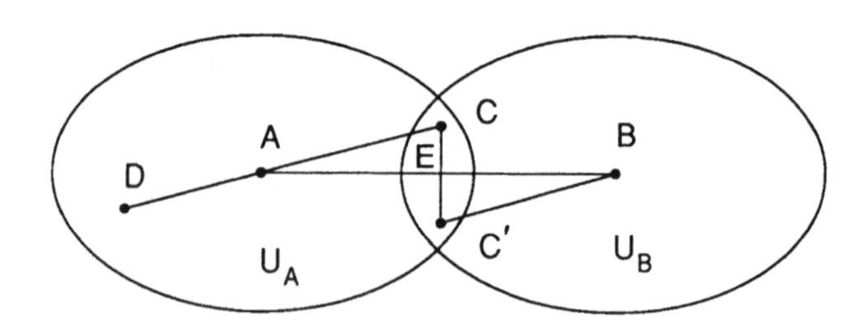

Fig. 27

Since $U_A\cap U_B=\emptyset$ for all $A,B\in L$, we have $\mathrm{vol}(U_0)<\Delta$. Hence $\mathrm{vol}(U)<2^n\Delta$, a contradiction. So we conclude that U contains at least one (and hence two) points besides 0. □

Recall that we have an embedding $A \to M(2,\mathbb{R})$:

$$x \to g_x = \begin{bmatrix} x_0 + x_1\sqrt{a} & x_2 + x_3\sqrt{a} \\ b(x_2 - x_3\sqrt{a}) & x_0 - x_1\sqrt{a} \end{bmatrix},$$

and that the quaternions $x \in \mathcal{O}$ are characterized by the conditions x_0, x_1, x_2, $x_3 \in \mathbb{Z}$.

First we show that for any $g \in SL(2,\mathbb{R})$, there exists g_x $(x \in \mathcal{O})$ such that $g_x g$ belongs to a fixed compact set of $M(2,\mathbb{R})$. Since A is a division algebra, $x \neq 0$ implies $\mathrm{Nrd}(x) = \det(g_x) \neq 0$; hence $g_x g \in GL(2,\mathbb{R})$. The norm on $GL(2,\mathbb{R})$ can be given as follows: any $h \in GL(2,\mathbb{R})$ can be written in the form $h = \lambda h_0$, where $\lambda = |\det h|^{1/2} > 0$ is a scalar matrix, and $\det(h_0) = \pm 1$. Suppose $h_0 = \begin{bmatrix} a & b \\ c & d \end{bmatrix}$. We define

$$||h|| = (a^2 + b^2 + c^2 + d^2)^{1/2} + |\ln \lambda|.$$

Let $g = \begin{bmatrix} \alpha & \beta \\ \gamma & \delta \end{bmatrix}$, $\alpha\delta - \beta\gamma = 1$. Then $g_x g = \begin{bmatrix} l_{11} & l_{12} \\ l_{21} & l_{22} \end{bmatrix}$, where

$$\begin{aligned} l_{11} &= \alpha x_0 + \sqrt{a}\alpha x_1 + \gamma x_2 + \sqrt{a}\gamma x_3, \\ l_{12} &= \beta x_0 + \sqrt{a}\beta x_1 + \delta x_2 + \sqrt{a}\delta x_3, \\ l_{21} &= \gamma x_0 - \sqrt{a}\gamma x_1 + b\alpha x_2 - b\sqrt{a}\alpha, \\ l_{22} &= \delta x_0 - \sqrt{a}\delta x_1 + b\beta x_2 - b\sqrt{a}\beta x_3. \end{aligned} \tag{5.4.1}$$

The entries of $g_x g$ are linear forms in x_0, x_1, x_2, x_3, and thus $(l_{11}, l_{12}, l_{21}, l_{22})$ can be considered as a lattice in $\mathbb{R}^4$ with determinant of the system (5.4.1) equal to 4ab.

Now we fix positive constants c_{11}, c_{12}, c_{21}, c_{22} such that their product is equal to 4ab, and consider a domain $\mathfrak{D} \subset \mathbb{R}^4$ given by $|l_{ij}| \le c_{ij}$. $\mathfrak{D}$ is a convex domain symmetric with respect to 0 of volume

$2^4 4ab$. By Lemma 5.4.2 we have $x_0, x_1, x_2, x_3 \in \mathbb{Z}$ such that $g_x g \in \mathfrak{D}$. We notice that $\mathfrak{D}$ corresponds to a set of real matrices with $|g_{ij}| < c_{ij}$ which is not compact. We are going to prove that, in fact, $g_x g$ actually belongs to a compact subset of $\mathfrak{D}$. For $m \neq 0$, we let

$$\mathfrak{D}_m = \{g \in \mathfrak{D} \mid \det(g) = m\}.$$

Each $\mathfrak{D}_m$ is compact; and since for $g \in \mathfrak{D}$, $|\det(g)| = |g_{11}g_{22} - g_{12}g_{21}| < K$ for some K, for large $|m|$, $\mathfrak{D}_m = \emptyset$. Since $x \in \mathcal{O}$, we have $\det(g_x g) = \det(g_x) = \mathrm{Nrd}(x) \in \mathbb{Z} - \{0\}$. Thus $g_x g \in K = \bigcup_{m \neq 0} \mathfrak{D}_m$, a compact subset of $\mathfrak{D}$. To complete the proof, we need the following lemma.

LEMMA 5.4.3. *Let* $x, y \in \mathcal{O}$. *They are called equivalent if* $x^{-1}y \in \mathcal{O}^1$. *Then the set of all quaternions from* $\mathcal{O}$ *with norm* m *consists of a finite number of classes of equivalent quaternions.*

PROOF of LEMMA: To each $x \in \mathcal{O}$, we associate a_x, the 4×4 matrix of the transformation $y \to yx$ written in the basis $\{1, i, j, k\}$. A direct calculation (compute $1 \cdot x$, $i \cdot x$, $j \cdot x$, $k \cdot x$ and put them into columns of a_x) shows that a_x has integral entries, and that if $\mathrm{Nrd}(x) = m$ then $\det(a_x) = m^2$. Let

$$GL_\Delta(n, \mathbb{Z}) = \{a \in GL(n, \mathbb{Z}) \mid \det(a) = \Delta\}.$$

There exists a finite set of matrices $a_1, \ldots, a_p \in GL_\Delta(n, \mathbb{Z})$ such that any matrix $a \in GL_\Delta(n, \mathbb{Z})$ can be written in the form

$$a = a_k \alpha \quad (\alpha \in SL(n, \mathbb{Z})).$$

This follows from Exercise 5.9. Thus among matrices a_x $(\mathrm{Nrd}(x) = m)$, there exist a finite number of matrices a_{x_i} $(\mathrm{Nrd}(x_i) = m)$ such that any matrix a_x $(\mathrm{Nrd}(x) = m)$ can be written as $a_x = a_{x_i}\alpha$, where $\alpha \in SL(n, \mathbb{Z})$. By Exercise 5.8, $\alpha = a_{x_i}^{-1} a_x = a_{x_i^{-1}x}$. Since

$\alpha \in SL(n,\mathbb{Z})$, $x_i^{-1}x \in \mathcal{O}^1$. □

We have $g_x g \in K$, a fixed compact subset of $GL(2, \mathbb{R})$. Choose x_i as in Lemma 5.4.3. Then $g_{x_i^{-1}} g_x = g_{x_i^{-1}x} \in \mathcal{O}^1$, and $g_{x_i^{-1}x} g \in g_{x_i^{-1}}(K) \subset \bigcup_{k=1}^{p} g_{x_k^{-1}}(K)$, a compact subset of $SL(2,\mathbb{R})$. The theorem follows from Corollary 3.6.2. □

5.5. The modular group and its subgroups

In this section we are going to study some important classes of subgroups of finite index in $\Gamma = PSL(2,\mathbb{Z})$.

Let n be a natural number, $n>1$, and let $\mathbb{Z}_n = \mathbb{Z}/n\mathbb{Z}$ be the ring of integers mod n. We are going to denote a congruence class which contains a given integer a by [a].

Let $SL(2,\mathbb{Z}_n)$ be the group of all unimodular matrices

$$[T] = \begin{bmatrix} [a] & [b] \\ [c] & [d] \end{bmatrix}$$

with entries from $\mathbb{Z}_n$. There is a natural homomorphism

$$\psi_n : \begin{bmatrix} a & b \\ c & d \end{bmatrix} \to \begin{bmatrix} [a] & [b] \\ [c] & [d] \end{bmatrix} \qquad (5.5.1)$$

of the group $SL(2,\mathbb{Z})$ to $SL(2,\mathbb{Z}_n)$. This in turn induces a homomorphism φ_n from $\Gamma = PSL(2,\mathbb{Z}) = SL(2,\mathbb{Z})/\{\pm 1_2\}$ to $PSL(2,\mathbb{Z}_n) = SL(2,\mathbb{Z}_n)/\{\pm 1_2\}$. The kernel of the homomorphism φ_n, $\Gamma(n)$, is called the *principal congruence subgroup of level* n. Obviously, the group $\Gamma(n)$ consists of those transformations $z \to \frac{az+b}{cz+d}$ for which $a \equiv d \equiv \pm 1 \pmod{n}$, $b \equiv c \equiv 0 \pmod{n}$.

THEOREM 5.5.1. *The map* ψ_n *is a surjective homomorphism of* $SL(2,\mathbb{Z})$ *onto* $SL(2,\mathbb{Z}_n)$.

PROOF: Let $[T]=\begin{bmatrix} [a] & [b] \\ [c] & [d] \end{bmatrix}$ be any matrix from $SL(2,\mathbb{Z}_n)$, and let a, b, c, d be arbitrary elements chosen from the corresponding congruence classes [a], [b], [c], [d]. Then $ad-bc\equiv 1 \pmod n$, i.e. $ad-bc=1+mn$ $(m\in\mathbb{Z})$. It follows that $(c,d)=\text{g.c.d.}(c,d)$ and n are relatively prime. Using Exercise 5.10, we may assume then that $(c,d)=1$. Let us consider a matrix

$$T=\begin{bmatrix} a+rn & b+sn \\ c & d \end{bmatrix}.$$

Its determinant is equal to $ad-bc+n(rd-sc)=1+n(m-rd-sc)$. Since d and c are relatively prime, we can choose integers r and s so that $m-rd-sc=0$, i.e. so that the matrix T is unimodular. Therefore we have proved that each matrix $[T]\in SL(2,\mathbb{Z}_n)$ has a preimage in $SL(2,\mathbb{Z})$. □

Let K(n) be the subgroup of $SL(2,\mathbb{Z})$ consisting of all unimodular matrices for which $\begin{bmatrix} a & b \\ c & d \end{bmatrix}\equiv\begin{bmatrix} 1 & 0 \\ 0 & 1 \end{bmatrix} \pmod n$. Then $K(n)=\text{Ker}\,\psi_n$.

COROLLARY 5.5.2.
(i) $SL(2,\mathbb{Z})/K(n)\approx SL(2,\mathbb{Z}_n)$, $\Gamma/\Gamma(n)\approx PSL(2,\mathbb{Z}_n)$;
(ii) $SL(2,\mathbb{Z}_2)=PSL(2,\mathbb{Z}_2)$, $|PSL(2,\mathbb{Z}_n)|=\frac{1}{2}|SL(2,\mathbb{Z}_n)|$ *for* $n>2$.

PROOF: Part (i) follows from Theorem 5.5.1 immediately. Since for $A\in SL(2,\mathbb{Z}_2)$ $(-1_2)A=A$ we have $SL(2,\mathbb{Z}_2)=PSL(2,\mathbb{Z}_2)$. In this case, the matrices in K(2) are precisely those corresponding to the transformations in $\Gamma(2)$. If $n>2$, $\begin{bmatrix} -1 & 0 \\ 0 & -1 \end{bmatrix} \notin K(n)$, and the matrices in $\{K(n)\cup(-I)K(n)\}$ correspond to the transformations in $\Gamma(n)$. Part (ii) follows. □

Now we are going to calculate the index $|\Gamma:\Gamma(n)|$, or equivalently, the order $|PSL(2,\mathbb{Z}_n)| = \frac{1}{2}|SL(2,\mathbb{Z}_n)|$, for $n>2$. We shall show that

$$|SL(2,\mathbb{Z}_n)| = n^3 \prod_{p|n} (1-\frac{1}{p^2}), \tag{5.5.2}$$

where the product is taken over all distinct primes dividing n.

Let p be a prime dividing n. Then there exists a natural homomorphism $\mathbb{Z}_n \to \mathbb{Z}_{\frac{n}{p}}$ of the ring of integers mod n to the ring of integers mod $\frac{n}{p}$. This homomorphism induces a homomorphism of the corresponding groups

$$SL(2,\ \mathbb{Z}_n) \to SL(2,\ \mathbb{Z}_{n/p}).$$

Let $I_{n,p}$ be the kernel of this homomorphism. Then we have

$$|SL(2,\ \mathbb{Z}_n)| = |I_{n,p}|\ |SL(2,\ \mathbb{Z}_{n/p})|.$$

We shall find $|I_{n,p}|$ using induction on the number of primes dividing n, in order to calculate $|SL(2,\mathbb{Z}_n)|$.

To calculate the order of the group $I_{n,p}$ we notice that it consists of all matrices $[T]\in SL(2,\ \mathbb{Z}_n)$ of the form

$$[T] = I + \frac{n}{p}[T_1].$$

In other words, $I_{n,p}$ consists of matrices

$$\begin{bmatrix} 1+\frac{n}{p}a & \frac{n}{p}b \\ \frac{n}{p}c & 1+\frac{n}{p}d \end{bmatrix},$$

where a, b, c, d $\in \mathbb{Z}_p$. By definition, the determinant of such a matrix is equal to 1 (mod n), i.e.

$$1+\frac{n}{p}(a+d)+\frac{n^2}{p^2}(ad-bc)\equiv 1 \pmod{n}.$$

Subtracting 1 and dividing by $\frac{n}{p}$ we get

$$a+d+\frac{n}{p}(ad-bc)\equiv 0 \pmod{p}. \tag{5.5.3}$$

The order of $I_{n,p}$ is equal to the number of solutions of this equation.

Case 1. $\frac{n}{p}$ is divisible by p. In this case (5.5.3) is equivalent to the equation

$$a+d\equiv 0 \pmod{p}. \tag{5.5.4}$$

Thus a, b, and c can be any integers mod p, and d is uniquely determined by (5.5.4). Thus, in this case $|I_{n,p}|=p^3$.

Case 2. Numbers p and $\frac{n}{p}$ are relatively prime. We can rewrite (5.5.3) as follows:

$$a(1+\tfrac{n}{p}d)+d-\tfrac{n}{p}bc\equiv 0 \pmod{p}. \tag{5.5.5}$$

Since $\left(\frac{n}{p}, p\right)=1$, $\frac{n}{p}d$ takes p different values mod p as $d\in\mathbb{Z}_p$. Therefore there are $(p-1)$ values that d can take to satisfy

$$1+\tfrac{n}{p}d\not\equiv 0 \pmod{p}, \tag{5.5.6}$$

b and c can be any integers mod p, and a is uniquely determined by b, c, and d. Thus the number of elements of $I_{n,p}$ satisfying (5.5.6) is equal to $p^2(p-1)$. There is a unique value of d to satisfy

$$1+\tfrac{n}{p}d\equiv 0 \pmod{p}; \tag{5.5.7}$$

bc takes a fixed value $\neq 0$ (mod p) determined by (5.5.5), and a is arbitrary. The number of elements satisfying (5.5.7) is equal then to $(p-1)p$. Thus, in this case

$$|I_{n,p}|=(p-1)p^2+(p-1)p=p^3(1-\frac{1}{p^2}).$$

Finally, we obtain the following formula:

$$I_{n,p}=\begin{cases} p^3, & \text{if p divides } \frac{n}{p} \\ p^3(1-\frac{1}{p^2}), & \text{if p does not divide } \frac{n}{p}. \end{cases}$$

In order to apply induction on the number of distinct prime divisors of n, we need to derive a formula for $|SL(2,\mathbb{Z}_p)|$ for prime p.

LEMMA 5.5.3. $|SL(2,\mathbb{Z}_p)|= p^3(1-\frac{1}{p^2})$.

PROOF: First we determine the order of $GL(2,\mathbb{Z}_p)$. The elements of this group are the matrices $\begin{bmatrix} [a] & [b] \\ [c] & [d] \end{bmatrix}$, with $[a]$, $[b]$, $[c]$, $[d]\in\mathbb{Z}_p$, such that the row vectors $([a],[b])$ and $([c],[d])$ are linearly independent. There are p^2-1 choices for $([a],[b])$, excluding $([0],[0])$; and for each $([a],[b])$ there are p^2-p choices for $([c],[d])$, excluding the p multiples of $([a],[b])$: so we have $|GL(2,\mathbb{Z}_p)|=(p^2-1)(p^2-p)$. Now $SL(2, \mathbb{Z}_p)$ is the kernel of a surjective homomorphism $\det : GL(2, \mathbb{Z}_p)\to\mathbb{Z}_p-\{0\}$, so $|SL(2, \mathbb{Z}_p)|= |GL(2, \mathbb{Z}_p)|/(p-1)=p(p^2-1)=p^3(1-\frac{1}{p^2})$. □

The formula (5.5.2) now follows by induction. Since $|PSL(2,\mathbb{Z}_n)|=\frac{1}{2}|SL(2, \mathbb{Z}_n)|$ for $n\neq 2$, and $|PSL(2, \mathbb{Z}_2)|=|SL(2, \mathbb{Z}_2)|=6$, we have proved the following theorem.

THEOREM 5.5.4.

$$|\Gamma : \Gamma(n)| = \begin{cases} 6, & \textit{if } n=2 \\ \frac{n^3}{2}\prod_{p|n}(1-\frac{1}{p^2}), & \textit{if } n>2, \end{cases}$$

where p *ranges over the distinct primes dividing* n. □

This result can be used to compute the hyperbolic area of the fundamental region for the group $\Gamma(n)$.

COROLLARY 5.5.5. *Let* F *be a fundamental region for* $\Gamma(n)$. *Then* $\mu(F)=\frac{\pi n^3}{6}\prod_{p|n}(1-\frac{1}{p^2})$, *if* $n>2$, *and* $\mu(F)=2\pi$, *if* $n=2$.

PROOF: Follows immediately from Theorem 5.5.4. □

Let us consider one more class of congruence subgroups of $SL(2,\mathbb{Z})$. Let K_n be a set of matrices in $SL(2, \mathbb{Z})$ of the form

$$\begin{bmatrix} a & nb \\ nc & d \end{bmatrix},$$

where a, b, c, d $\in\mathbb{Z}$. Obviously, $K_n \supset K(n)$. Let Γ_n be a subgroup of Γ, where $\Gamma_n=\left\{z\rightarrow\frac{az+b}{cz+d} \mid b\equiv c\equiv 0 \pmod n\right\}$. We have $\Gamma_n\supset\Gamma(n)$. Since $(-1_2)\in K_n$, we have $SL(2, \mathbb{Z})/K_n \approx \Gamma/\Gamma_n$. Hence $|\Gamma:\Gamma_n|=|SL(2, \mathbb{Z}):K_n|$. By Exercise 5.11, $|K_n : K(n)|=\varphi(n)=n\prod_{p|n}(1-\frac{1}{p})$, where $\varphi(n)$ is the *Euler function* equal to the number of natural numbers $x<n$, relatively prime to n. Hence we have

$$|\Gamma:\Gamma_n|=|SL(2, \mathbb{Z}) : K_n|=\frac{|SL(2, \mathbb{Z}) : K(n)|}{|K_n : K(n)|}=\frac{n^3\prod(1-\frac{1}{p^2})}{n\prod(1-\frac{1}{p})}=n^2\prod_{p|n}(1+\frac{1}{p}).$$

If F is a fundamental region for Γ_n then we have $\mu(F)=\frac{\pi n^2}{3}\prod_{p|n}(1+\frac{1}{p})$.

Our next task is to prove that any subgroup Λ of finite index in $\Gamma=PSL(2,\mathbb{Z})$ has finitely many (≥ 1) conjugacy classes of maximal parabolic subgroups, and hence that $\Lambda\backslash\mathcal{H}$ is not compact and has the same number of cusps (see §4.2). This will imply that the parabolic class number of Λ is finite and ≥ 1. This result will certainly be valid for the congruence subgroups considered above.

LEMMA 5.5.6. $\Gamma=PSL(2,\mathbb{Z})$ acts transitively on $\hat{\mathbb{Q}}=\mathbb{Q}\cup\{\infty\}$.

PROOF: For any $\frac{a}{c}\in\mathbb{Q}$ such that $(a,c)=1$, there exist $d,b\in\mathbb{Z}$ such that $ad-bc=1$, and therefore a transformation $T(z)=\frac{az+b}{cz+d}$ which maps ∞ to $\frac{a}{c}$. Thus any two points in $\hat{\mathbb{Q}}$ are congruent under Γ. □

LEMMA 5.5.7. *For any* $r\in\hat{\mathbb{Q}}$ *there exists a maximal parabolic subgroup* $\Gamma_r\leq\Gamma$ *stabilizing* r.

PROOF: Using Lemma 5.5.6, we construct $T\in\Gamma$ mapping r to ∞. Let Γ_∞ be a cyclic subgroup generated by Z: $z\to z+1$. It is a maximal parabolic subgroup of Γ. Then $\Gamma_r=T^{-1}\Gamma_\infty T$ is a maximal parabolic subgroup of Γ fixing r. □

Now we need the following general result.

LEMMA 5.5.8. *If* A *and* B *are subgroups of a group* G, *and* $C=A\cap B$, *then* $|B:C|\leq|G:A|$.

PROOF: To each coset bC of C in B, we associate the coset bA of A in G. This does not depend on the choice of coset representative, since if $b_1C=b_2C$, then $b_1b_2^{-1}\in C\leq A$ implies $b_1A=b_2A$. Distinct cosets bC correspond to distinct cosets bA: for if $b_1A=b_2A$ (with b_1, $b_2\in B$), then $b_1b_2^{-1}\in A\cap B=C$ giving $b_1C=b_2C$. Thus there are at least as many cosets of A in G as there are of C in B. □

COROLLARY 5.5.9. *$\Lambda_r = \Lambda \cap \Gamma_r$ is a maximal parabolic subgroup for each $r \in \hat{\mathbb{Q}}$, and any parabolic subgroup in Λ can be obtained this way.*

PROOF: Apply Lemma 5.5.8 with $G=\Gamma$, $A=\Lambda$, $B=\Gamma_r$, $C=\Lambda \cap \Gamma_r = \Lambda_r$. We have $|\Gamma_r : \Lambda_r| \leq |\Gamma : \Lambda|$, which is finite. Thus Λ_r is infinite, and therefore non-trivial for each $r \in \hat{\mathbb{Q}}$. Λ_r fixes r, since Γ_r does (Lemma 5.5.7). Suppose Λ_r is contained in a larger cyclic subgroup. By Theorem 2.3.2, however, this larger subgroup also fixes r, and therefore is contained in Γ_r, which is maximal and in Λ_r; so Λ_r is a maximal parabolic subgroup for each $r \in \hat{\mathbb{Q}}$. Now suppose C is a maximal parabolic subgroup in Λ. C has a unique fixed point $r \in \mathbb{R} \cup \infty$. Solving $cr^2 + (d-a)r - b = 0$ ($a, b, c, d \in \mathbb{Z}$, $ad-bc=1$, $|a+d|=2$), we see that $r = \frac{a-d}{2c} \in \hat{\mathbb{Q}}$. Hence $C \leq \Lambda_r$; and since C is maximal, $C = \Lambda_r$. □

LEMMA 5.5.10. *The conjugacy classes of maximal parabolic subgroups in Λ are in one-to-one correspondence with Λ-orbits on $\hat{\mathbb{Q}}$.*

PROOF: Let $r_1, r_2 \in \hat{\mathbb{Q}}$, and $\Lambda_{r_1}, \Lambda_{r_2}$ be their stabilizers in Λ. If $r_2 = Tr_1$ then $\Lambda_{r_1} = T^{-1}\Lambda_{r_2}T$. Conversely, suppose Λ_1 and Λ_2 are two maximal parabolic subgroups conjugate in Λ. We have seen in Corollary 5.5.9 that each parabolic subgroup has a unique fixed point $r \in \hat{\mathbb{Q}}$. Hence $\Lambda_1 = \Lambda_{r_1}$, $\Lambda_2 = \Lambda_{r_2}$, and we have

$$\Lambda_{r_1} = T^{-1}\Lambda_{r_2}T. \tag{5.5.8}$$

Applying both parts of (5.5.8) to r_1, we obtain $T(r_1) = \Lambda_{r_2}T(r_1)$ which implies $T(r_1) = r_2$. □

COROLLARY 5.5.11. *The parabolic class number* s *of* Λ *satisfies* $1 \leq s \leq N$, *where* N *is the index* $|\Gamma : \Lambda|$; in particular, s *is finite.*

PROOF: Since s is the number of orbits of Λ on $\hat{\mathbb{Q}}$, we have $s \geq 1$. Let $\Lambda T_1, \ldots, \Lambda T_N$ be the cosets of Λ in Γ. For any $r \in \hat{\mathbb{Q}}$, we have $r = T(\infty)$ for some $T \in \Gamma$ (by Lemma 5.5.6). But $T = ST_i$, for some $S \in \Lambda$ and $i = 1, \ldots, N$. Thus $r = ST_i(\infty)$ lies in the Λ-orbit of $T_i(\infty)$, for some $1 \leq i \leq N$. So Λ has at most N orbits, which implies $s \leq N$. □

Thus we have proved (with the help of Theorem 4.2.1) the following theorem.

THEOREM 5.5.12. *Let* $\Lambda \leq \Gamma$ *be a subgroup in* Γ *of finite index* N. *Then* Λ *has a finite number* $1 \leq s \leq N$ *of conjugacy classes of parabolic subgroups, and therefore* $\Lambda \backslash \mathcal{H}$ *is not compact, but is of finite hyperbolic area, having* s *cusps.* □

EXAMPLE F. Consider $\Lambda = \Gamma(2)$, the principal congruence subgroup of level 2 and index 6 in Γ. Λ has three orbits on $\hat{\mathbb{Q}}$, namely

$[0]_\Lambda = \{p/q \mid p, q \in \mathbb{Z}, \text{ p is even, q is odd}\}$,

$[1]_\Lambda = \{p/q \mid p, q \in \mathbb{Z}, \text{ p and q are odd}\}$,

$[\infty]_\Lambda = \{p/q \mid p, q \in \mathbb{Z}, \text{ p is odd and q is even}\} \cup \{\infty\}$,

represented by verteces at infinity $r_1 = 0$, $r_2 = 1$, $r_3 = \infty$. As coset representatives for Λ in Γ we can take the elements

$$\{I, X, Z, ZY, ZX, ZYX\} = \{T_1, T_2, \ldots, T_6\}$$

corresponding to the matrices

$$\begin{bmatrix} 1 & 0 \\ 0 & 1 \end{bmatrix}, \begin{bmatrix} 0 & 1 \\ -1 & 0 \end{bmatrix}, \begin{bmatrix} 1 & 1 \\ 0 & 1 \end{bmatrix},$$

$$\begin{bmatrix} 1 & 0 \\ 1 & 1 \end{bmatrix}, \begin{bmatrix} -1 & 1 \\ -1 & 0 \end{bmatrix}, \begin{bmatrix} 0 & 1 \\ -1 & 1 \end{bmatrix}$$

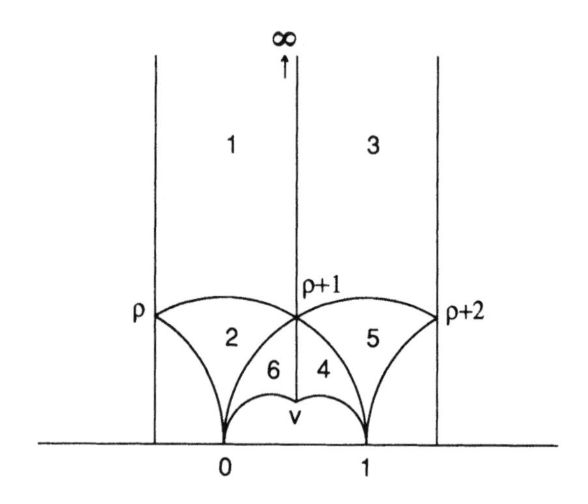

Fig. 28

in SL(2,$\mathbb{Z}$). (X, Y, Z are defined in §3.6 (see Fig. 11)). The fundamental region $G = \bigcup_{j=1}^{6} T_j(F)$ for Λ is illustrated in Figure 28 where the index j denotes the region $T_j(F)$ (j=1, ... ,6). G is bounded by six geodesics; a pair of sides meeting at each of three vertices at infinity are paired by elements of Λ (e.g. the two vertical sides are congruent under Z^2), and after the identification we obtain a sphere with three cusps. Notice that the three vertices ρ, $\rho+2$ and $v=(\rho+1)/(\rho+2)=\frac{1}{2}+i/2\sqrt{3}$ are congruent under Λ, so they are mapped to a single point in $\Lambda\backslash\mathcal{H}$. The sum of the angles at these three points is equal to 2π, so it is not a marked point.

5.6. Examples

Here we give various examples of arithmetic and non-arithmetic groups.

EXAMPLE G. [K] Let $A=\left(\frac{3,5}{\mathbb{Q}}\right)$. Since 3 is a quadratic non-residue (mod 5), A is a division algebra over $\mathbb{Q}$ (Theorem 5.2.5). Let us consider a $\mathbb{Z}$-module $\mathcal{O}$ generated by $\{1, i, \frac{1+j}{2}, \frac{i+k}{2}\}$. By Exercise 5.13, $\mathcal{O}$ is an order. In fact, $\mathcal{O}$ is a maximal order, i.e. it is not contained in any other order. The group of units of norm 1 in $\mathcal{O}$, Γ, is embedded into PSL(2,$\mathbb{R}$) as follows (a slightly different embedding than in (5.2.4):

$$\Gamma \to \left\{\pm\begin{bmatrix} \frac{l+m\sqrt{3}}{2} & \sqrt{5}\left(\frac{w-u\sqrt{3}}{2}\right) \\ \sqrt{5}\left(\frac{w+u\sqrt{3}}{2}\right) & \frac{l-m\sqrt{3}}{2} \end{bmatrix}\right.$$

$$\left| \; (l,m,u,w)\in\mathbb{Z}^4,\ l\equiv w \pmod 2,\ m\equiv u \pmod 2,\ l^2-3m^2-5w^2+15u^2=4\right\}.$$

The group $\Gamma_0=R\Gamma R^{-1}$, where $R=\begin{bmatrix} i & 1 \\ 1 & i \end{bmatrix}$ acts on the unit disc $\mathcal{U}$.

Let us denote $RTR^{-1}=\begin{bmatrix} a & \bar{c} \\ c & \bar{a} \end{bmatrix}$. Then $a=\frac{l-iu\sqrt{15}}{2}$, $c=\frac{w\sqrt{5}-im\sqrt{3}}{2}$,

$|a|^2=\frac{r}{4}+1$, and $|c|^2=\frac{r}{4}$, so $|a|^2$, and $|c|^2\in\frac{1}{4}\mathbb{Z}$.

We explain now how to construct the Ford fundamental region R_0, which is compact by Theorem 5.4.1. We saw in the proof of Theorem 3.3.7 that given any $A>0$ there are only finitely many elements of Γ_0 with $|a|<A$. This follows from the relation $|a|^2-|c|^2=1$ and the discreteness of Γ_0. We can thus list all

elements of Γ_0 in order of increasing $|a|$. This list will eventually include all elements of Γ_0. Taking isometric circles for the elements according to their order, we shall obtain the fundamental region R_0 after a finite number of steps. Indeed, the distance from the isometric circle $I(T)$, $T=\begin{bmatrix} a & \bar{c} \\ c & \bar{a} \end{bmatrix}$, to the center of $\mathcal{U}$ is equal to $\frac{|a|-1}{|c|}$ which tends to 1 as $|a|\to\infty$. Thus isometric circles with sufficiently large $|a|$ cannot contribute to the boundary of the compact fundamental region.

In Table 1, we give the beginning of the list of elements of Γ. (For $r=0$ we get the identity element, which we do not include in the table). Columns 1 – 5 give values of r, l, m, u, w. Columns 6 – 8 give the coordinates (x,y) of the center of the corresponding isometric circle and its radius R. The isometric circles of the first 8 elements form a boundary of the fundamental region R_0 (see Fig. 29), and therefore can be chosen as generators of the group Γ_0. The finite part of the tessellation of $\mathcal{U}$ by the images of R_0 is given in Figure 29.

Here we list the generators of the group Γ_0:

$$T_1=\begin{bmatrix} \frac{3}{2} & -\frac{\sqrt{5}}{2} \\ -\frac{\sqrt{5}}{2} & \frac{3}{2} \end{bmatrix}, \quad T_2=\begin{bmatrix} 2-\sqrt{3} & 0 \\ 0 & 2+\sqrt{3} \end{bmatrix},$$

$$T_3=\begin{bmatrix} \frac{4-3\sqrt{3}}{2} & \frac{\sqrt{15}}{2} \\ -\frac{\sqrt{15}}{2} & \frac{4+3\sqrt{3}}{2} \end{bmatrix},$$

$$T_4 = \begin{bmatrix} \frac{4+3\sqrt{3}}{2} & \frac{\sqrt{15}}{2} \\ -\frac{\sqrt{15}}{2} & \frac{4-3\sqrt{3}}{2} \end{bmatrix}.$$

r	l	m	u	w	x	y	R
5	3	0	0	−1	1.342	0.000	0.894
5	3	0	0	1	−1.342	0.000	0.894
12	4	−2	0	0	0.000	1.155	0.577
12	4	2	0	0	0.000	−1.555	0.577
27	4	−3	−1	0	0.745	0.770	0.385
27	4	3	−1	0	−0.745	−0.770	0.385
27	4	−3	1	0	−0.745	0.770	0.385
27	4	3	1	0	0.745	−0.770	0.385
32	6	−2	0	−2	0.839	0.650	0.354
32	6	−2	0	2	−0.839	0.650	0.354
32	6	2	0	−2	0.839	−0.650	0.354
32	6	2	0	2	−0.839	−0.650	0.354
45	7	0	0	−3	1.043	0.000	0.298
45	7	0	0	3	−1.043	0.000	0.298
47	6	−3	−1	−2	0.999	0.295	0.292
47	6	−3	−1	2	−0.143	1.032	0.292
47	6	3	−1	−2	0.143	−1.032	0.292
47	6	3	−1	2	−0.999	−0.295	0.292
47	6	−3	1	−2	0.143	1.032	0.292
47	6	−3	1	2	−0.999	0.295	0.292
47	6	3	1	−2	0.999	−0.295	0.292
47	6	3	1	2	−0.143	−1.032	0.292
57	1	−2	−2	−3	0.588	−0.851	0.265
57	1	−2	−2	3	0.353	0.972	0.265
57	1	2	−2	−3	−0.353	−0.972	0.265
57	1	2	−2	3	−0.588	0.851	0.265
57	1	−2	2	−3	−0.353	0.972	0.265
57	1	−2	2	3	−0.588	−0.851	0.265
57	1	2	2	−3	0.588	0.851	0.265
57	1	2	2	3	0.353	−0.972	0.265

Table 1

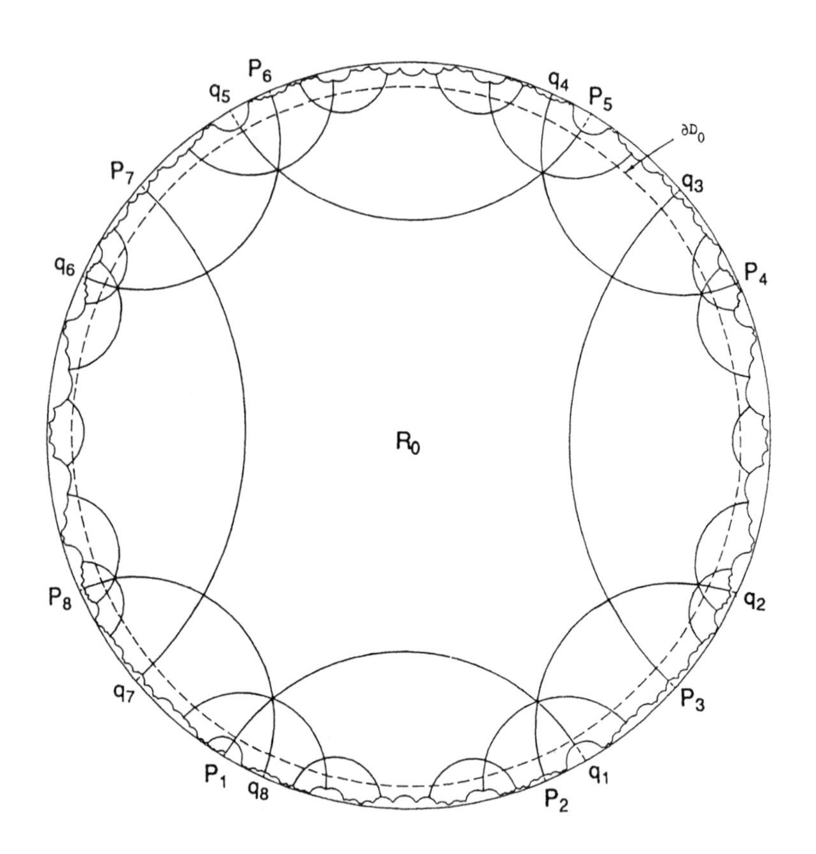

Fig. 29

The identifications of the sides of the fundamental region R_0 are shown in Figure 30. The genus of $\Gamma_0\backslash\mathcal{H}$ is 1, and the number of non-congruent elliptic points of order 3 is equal to 2. By Theorem 4.3.1, $\mu(R_0)=\frac{8\pi}{3}$.

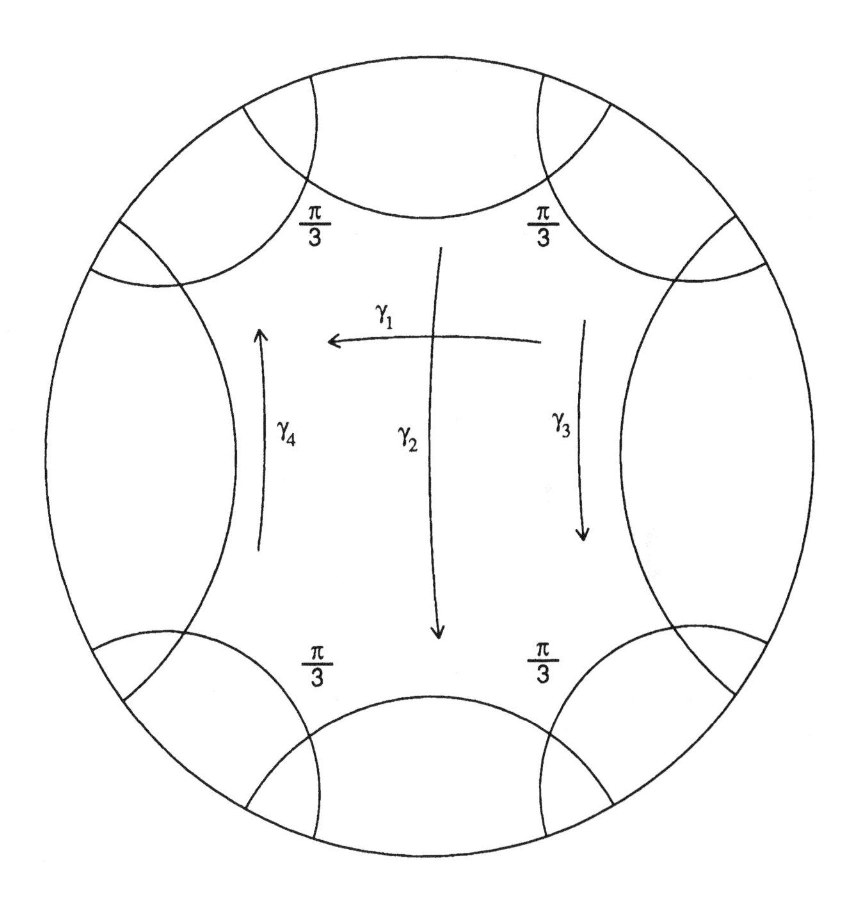

Fig. 30

EXAMPLE C. Let us go back to the example of the Fuchsian group Γ_8 of signature $(2\ ;\ -)$ constructed in § 4.3. Its generators are:

$$A_2=\begin{bmatrix} \dfrac{(2+\sqrt{2})(1-\sqrt[4]{2})}{2} & \dfrac{(2+\sqrt{2})-\sqrt[4]{2}\sqrt{2}}{2} \\[2em] -\dfrac{(2+\sqrt{2})+\sqrt[4]{2}\sqrt{2}}{2} & \dfrac{(2+\sqrt{2})(1+\sqrt[4]{2})}{2} \end{bmatrix},$$

$$A_1=\begin{bmatrix} \dfrac{(2+\sqrt{2})(1+\sqrt[4]{2})}{2} & \dfrac{(2+\sqrt{2})+\sqrt[4]{2}\sqrt{2}}{2} \\[2em] -\dfrac{(2+\sqrt{2})-\sqrt[4]{2}\sqrt{2}}{2} & \dfrac{(2+\sqrt{2})(1-\sqrt[4]{2})}{2} \end{bmatrix},$$

$$B_1=\begin{bmatrix} \dfrac{(2+\sqrt{2})(1-\sqrt[4]{2})}{2} & \dfrac{-(2+\sqrt{2})+\sqrt[4]{2}\sqrt{2}}{2} \\[2em] -\dfrac{-(2+\sqrt{2})-\sqrt[4]{2}\sqrt{2}}{2} & \dfrac{(2+\sqrt{2})(1+\sqrt[4]{2})}{2} \end{bmatrix},$$

$$B_2=\begin{bmatrix} \frac{(2+\sqrt{2})(1+\sqrt[4]{2})}{2} & \frac{-(2+\sqrt{2})-\sqrt[4]{2}\sqrt{2}}{2} \\ -\frac{-(2+\sqrt{2})+\sqrt[4]{2}\sqrt{2}}{2} & \frac{(2+\sqrt{2})(1-\sqrt[4]{2})}{2} \end{bmatrix}.$$

Using Exercise 5.15, we conclude that $tr(\Gamma_8)$ in contained in the ring of integers of $\mathbf{Q}(\sqrt{2})$. $\mathbf{Q}(\sqrt{2})$ is a totally real quadratic extension of $\mathbf{Q}$, so condition (I) of Theorem 5.3.4 is satisfied.

Now let φ_2: $\mathbf{Q}(\sqrt{2}) \to \mathbf{R}$ be a non-identity embedding sending $\sqrt{2} \to -\sqrt{2}$. Notice that all generators of Γ_8, and hence all elements of Γ_8, are embedded into M(2,K) where $K=\mathbf{Q}(\sqrt{2})(\sqrt{\sqrt{2}})$. Then φ_2 extends to an isomorphism ψ_2: K $\to$ $\mathbf{C}$, such that

$$\sqrt[4]{2} \to \sqrt{-\sqrt{2}} = i\cdot\sqrt[4]{2}.$$

Under this embedding the generators are mapped into the following matrices in M(2,$\mathbf{C}$):

$$A'_2=\begin{bmatrix} \frac{(2-\sqrt{2})(1-i\cdot\sqrt[4]{2})}{2} & \frac{(2-\sqrt{2})+i\cdot\sqrt[4]{2}\sqrt{2}}{2} \\ -\frac{(2-\sqrt{2})-i\cdot\sqrt[4]{2}\sqrt{2}}{2} & \frac{(2-\sqrt{2})(1+i\cdot\sqrt[4]{2})}{2} \end{bmatrix},$$

$$A'_1=\begin{bmatrix} \frac{(2-\sqrt{2})(1+i\cdot\sqrt[4]{2})}{2} & \frac{(2-\sqrt{2})-i\cdot\sqrt[4]{2}\sqrt{2}}{2} \\ -\frac{(2-\sqrt{2})+i\cdot\sqrt[4]{2}\sqrt{2}}{2} & \frac{(2-\sqrt{2})(1-i\cdot\sqrt[4]{2})}{2} \end{bmatrix},$$

$$B'_1=\begin{bmatrix} \frac{(2-\sqrt{2})(1-i\cdot\sqrt[4]{2})}{2} & \frac{-(2-\sqrt{2})-i\cdot\sqrt[4]{2}\sqrt{2}}{2} \\ -\frac{-(2-\sqrt{2})+i\cdot\sqrt[4]{2}\sqrt{2}}{2} & \frac{(2-\sqrt{2})(1+i\cdot\sqrt[4]{2})}{2} \end{bmatrix},$$

$$B'_2=\begin{bmatrix} \frac{(2-\sqrt{2})(1+i\cdot\sqrt[4]{2})}{2} & \frac{-(2-\sqrt{2})+i\cdot\sqrt[4]{2}\sqrt{2}}{2} \\ -\frac{-(2-\sqrt{2})-i\cdot\sqrt[4]{2}\sqrt{2}}{2} & \frac{(2-\sqrt{2})(1-i\cdot\sqrt[4]{2})}{2} \end{bmatrix}.$$

Thus

$$A^{\varphi_2}=\left\{\begin{bmatrix} a & b \\ -\bar{b} & \bar{a} \end{bmatrix} \mid a, b \in \psi_2(K)\right\}.$$

It follows from Exercise 5.5 that $A^{\varphi_2}\otimes\mathbf{R}\approx\mathbf{H}$. Hence by Example E (§5.2), $\varphi_2(\mathrm{tr}(\Gamma))$ are bounded in $\mathbf{C}$. So (II) is also satisfied, and by Theorem 5.3.4, Γ_8, is derived from a quaternion algebra over $\mathbf{Q}(\sqrt{2})$.

EXAMPLE H. Let Γ be a triangle group with signature $(0; \frac{\pi}{2}, \frac{\pi}{m}; 1)$. According to the construction in §4.4, we first generate a group Γ^* by hyperbolic reflections in the sides of the triangle with vertices v_1, $v_2=i$, $v_3=\infty$ and the angles at these vertices $\frac{\pi}{m}$, $\frac{\pi}{2}$, and 0, respectively: $R_1(z)=-\bar{z}$, $R_2(z)=-\bar{z}+2\cos\frac{\pi}{m}$, $R_3=\frac{1}{\bar{z}}$ (see Fig. 31). The triangle group Γ is generated by $R_1R_3=-\frac{1}{z}$ and $R_2R_1=z+2\cos\frac{\pi}{m}$ which identify two pairs of sides of the fundamental quadrangle v_1, v_2, v_1', v_3: R_1R_3 identifies $v_1'v_2$ with v_1v_2, and R_2R_1 identifies $v_1'v_3$ with v_1v_3. The corresponding group of matrices is called the *Hecke group* and is denoted $\Gamma(2\cos\frac{\pi}{m})$. Since $2\cos\frac{\pi}{m}$ is always an algebraic integer, $\Gamma(2\cos\frac{\pi}{m})$ is a group of matrices over the integers in $\mathbb{Q}(2\cos\frac{\pi}{m})$, but it is easy to see that not every matrix over the integers in $\mathbb{Q}(2\cos\frac{\pi}{m})$ belongs to this group. We are going to prove that $\Gamma(2\cos\frac{\pi}{m})$ is arithmetic if and only if m=3, 4 , or 6.

For m=3 we obtain $\Gamma(2\cos\frac{\pi}{3})=\Gamma(1)=SL(2,\mathbb{Z})$, an arithmetic Fuchsian group derived from the matrix quaternion algebra. For $m>3$, let $\lambda=2\cos\frac{\pi}{m}$, and $\Gamma(\lambda)$ be the subgroup of $SL(2,\mathbb{R})$ generated by the following two elements:

$$S=\begin{bmatrix} 0 & 1 \\ -1 & 0 \end{bmatrix}, \quad T_\lambda=\begin{bmatrix} 1 & \lambda \\ 0 & 1 \end{bmatrix}.$$

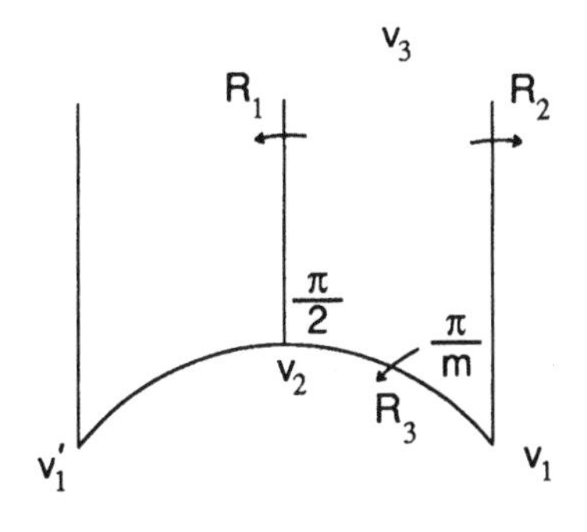

Fig. 31

According to Exercise 5.16, $k_\lambda=\mathbf{Q}(tr(T)\ |T\in\Gamma(\lambda))=\mathbf{Q}(\lambda)$ is a totally real number field of degree $\frac{1}{2}\varphi(2m)$, where φ is the Euler function. Since $\lambda\in\mathcal{O}_{k_\lambda}$, $\Gamma(\lambda)$ is a subgroup of $SL(2,\mathcal{O}_{k_\lambda})$. Therefore we have $tr(\Gamma(\lambda))\subset\mathcal{O}_{k_\lambda}$. The element $ST_\lambda{}^m$ belongs to $\Gamma(\lambda)$ for any $m\in\mathbb{Z}$, and since $tr(ST_\lambda{}^m)=-m\lambda$ for any embedding $\varphi_i : k_\lambda \to \mathbb{C}$, the set $\varphi_i(tr(\Gamma))$ is not bounded. Hence for $m>3$, $\Gamma(\lambda)$ is not derived from a quaternion algebra.

By Exercise 5.17, for m=4 or 6, $k_\lambda{}^{(2)}=\mathbf{Q}$, and hence $\Gamma(\lambda)^{(2)}$is derived from a quaternion algebra (since $\Gamma(\lambda)$ has parabolic elements this quaternion algebra must be the matrix algebra, see Theorems 4.2.1 and 5.4.1. Hence $\Gamma(\sqrt{2})$ and $\Gamma(\sqrt{3})$ are arithmetic Fuchsian groups. For m=5 or m≥7, $k_\lambda{}^{(2)}$ is a proper extension of $\mathbf{Q}$. Since $\Gamma(\lambda)^{(2)}$ contains $(ST_\lambda)^2T_\lambda{}^{2m}$ for all $m\in\mathbb{Z}$, and since $tr((ST_\lambda)^2T_\lambda{}^{2m})=(2m+1)\lambda^2-2$, for any embedding $\varphi_i : k_\lambda{}^{(2)}\to \mathbb{C}$, the set $\varphi_i(tr(\Gamma))$ is not bounded, hence $\Gamma(\lambda)$ is not arithmetic.

EXERCISES FOR CHAPTER 5

5.1. Prove that the map (5.2.4) satisfies $g_x+g_y=g_{x+y}$, and $g_x \cdot g_y=g_{xy}$.

5.2. Let $p\equiv-1 \pmod 4$ be a prime number. Prove that $A=\left(\frac{-1,p}{\mathbb{Q}}\right)$ is a division algebra.

5.3. Show that there are only two quaternion algebras over $\mathbb{R}$ up to isomorphism: $\left(\frac{a,b}{\mathbb{R}}\right)\approx\mathbb{H}$ if $a<0$, $b<0$, and $\left(\frac{a,b}{\mathbb{R}}\right)\approx M(2,\mathbb{R})$ otherwise.

5.4. Prove that the three definitions of an order given in §5.2 are equivalent.

5.5. Use Lemma 5.3.7 to show that $A_i\otimes\mathbb{R}\approx\mathbb{H}$.

5.6. Prove that $\Gamma/\Gamma^{(2)}$ is a finite abelian group of type $(2, 2, \cdots, 2)$, where Γ and $\Gamma^{(2)}$ are as at the end of §5.3.

5.7. Let Γ and $\Gamma^{(2)}$ be as above. Prove that $k_2=\mathbb{Q}(\mathrm{tr}(T)^2 \mid T\in\Gamma)$ coincides with $k_2'=\mathbb{Q}(\mathrm{tr}(T) \mid T\in\Gamma^{(2)})$.

5.8. Prove that the correspondence $x \to a_x$ associates to the product of quaternions the product of corresponding 4×4 matrices.

5.9. Each integral non-degenerate $n\times n$ matrix a can be multiplied by an appropriate integral unimodular matrix α in such a way that $a\alpha$ is a lower-triangular matrix with $|a_{ij}|<|a_{ii}|$ for $i>j$ $(i, j=1, \dots, n)$.

5.10. If (c,d) and n are relatively prime, then there exists q such that c and $d+qn$ are relatively prime.

5.11. Prove that $|K_n : K(n)|=\varphi(n)$.

5.12. Let $K_0(n)$ be a subgroup of $SL(2,\mathbb{Z})$ consisting of all matrices of the form

$$\begin{bmatrix} a & nb \\ c & d \end{bmatrix},$$

where $a, b, c, d\in\mathbb{Z}$. Prove that $|SL(2,\mathbb{Z}) : K_0(n)|=n\prod_{p|n}(1+\frac{1}{p})$.

5.13. Prove that $\mathcal{O}$ in Example G is an order of A.

5.14.[M] Let Γ be a finitely generated subgroup of $SL(2,\mathbb{R})$. Let $\{S_1, \dots, S_r\}$ be a set of generators of Γ. For any subset $\{i_1, \dots, i_s\} \subset \{1, \dots, r\}$ put $t_{i_1 \dots i_s} = tr(T_{i_1} \cdots T_{i_s})$. Then $tr(\Gamma)$ is contained in the ring $\mathbb{Z}[t_{i_1 \dots i_s} \mid \{i_1, \dots, i_s\} \subset \{1, \dots, r\}]$.

5.15.Let T_1, T_2, T_3, T_4 be generators of the group Γ_8 as given in §5.6 (Example C). Prove that for any subset $\{i_1, \dots, i_s\} \subset \{1, 2, 3, 4\}$

$tr(T_{i_1} \cdots T_{i_s}) \in \mathcal{O}_{\mathbb{Q}(\sqrt{2})}$.

5.16.Prove that $k_\lambda = \mathbb{Q}(tr(T) \mid T \in \Gamma(\lambda)) = \mathbb{Q}(\lambda)$ is a totally real number field of degree $\frac{1}{2}\varphi(2m)$.

5.17.Prove that $k_\lambda^{(2)} = \mathbb{Q}(tr(T) \mid T \in \Gamma(\lambda)^{(2)}) = \mathbb{Q}(\lambda^2)$.

HINTS FOR SELECTED EXERCISES

1.1. $T(z)=-\frac{1}{z-\alpha}+\beta=\frac{\beta z-(\alpha\beta+1)}{z-\alpha}$ is given by the matrix $\begin{bmatrix} \beta & -\alpha\beta-1 \\ 1 & -a \end{bmatrix}$ of determinant 1, hence $T\in PSL(2,\mathbb{R})$. Let α' be the second point where the geodesic meets $\mathbb{R}\cup\{\infty\}$. If $\alpha'=\infty$, then $\beta=0$. If $\alpha'\neq\infty$, then $\beta=(\alpha'-\alpha)^{-1}$.

1.2. It is obvious that the map $z\to-\bar{z}$ preserves the norm on $T\mathcal{H}$, hence it is sufficient to consider the case when $g(z)=\frac{az+b}{cz+d}\in PSL(2,\mathbb{R})$. Let $\zeta\in T_z\mathcal{H}$. Then $(Dg)(\zeta)=g'(z)\cdot\zeta$, and hence by (1.1.3) and (1.1.4)

$$\|(Dg)(\zeta)\|=\frac{|(Dg)(\zeta)|}{\operatorname{Im}(gz)}=\frac{|g'(z)||\zeta|}{\operatorname{Im}(gz)}=\frac{|\zeta|}{\operatorname{Im}z}=\|\zeta\|. \qquad (*)$$

Conversely, let $(*)$ hold, and γ: $I\to\mathcal{H}$ be a piecewise differentiable path in $\mathcal{H}$ given by $z(t)=(x(t),y(t))$. Then by $(*)$

$$h(g(\gamma))=\int_0^1\frac{|g'(z(t))|\cdot|z'(t)|dt}{\operatorname{Im}(g(z(t)))}=\int_0^1\frac{|z'(t)|dt}{\operatorname{Im}((z(t))}=h(\gamma).$$

1.3. Let $z=x+iy$. The inner product in $T_z\mathcal{H}$ (which can be identified with $\mathbb{R}^2$) given by the formula (1.3.3) is clearly a scalar multiple of the Euclidean inner product in $\mathbb{R}^2$. Use the formula $\cos\theta=\frac{<\zeta_1,\zeta_2>}{\|\zeta_1\|\cdot\|\zeta_2\|}$ to obtain the result.

1.7. Use Exercise 1.6 and a calculation similar to one in the proof of Theorem 1.2.1 to show that the geodesic joining 0 and ir is the segment of the diameter joining these points. Use the transformation (1.2.2) and Exercise 1.1 to show that any circle orthogonal to Σ and any diameter can be transformed to a vertical diameter via a transformation in PSL(2,$\mathbb{R}$).

1.9. Orientation-preserving isometries of $\mathcal{U}$ are transformations of the form: $f \circ T \circ f^{-1}$, where $T \in SL(2,\mathbb{R})$ and f is of form (1.2.2).

1.11. Use Theorem 1.2.6(v).

1.13. Using the $\mathcal{U}$ model and considereng a circle centered at 0 of hyperbolic radius r, prove that in the hyperbolic case, $\odot r = 2\pi \sinh r$ (you shall need formulae from Exercise 1.8).

2.2. Use Theorem 1.2.6(ii) to obtain

$$\cosh \rho(i, T(i)) = \tfrac{1}{2}(a^2+b^2+c^2+d^2).$$

2.3. Let $g = \begin{bmatrix} \alpha & \beta \\ \gamma & \delta \end{bmatrix} \in PSL(2,\mathbb{R})$, $z = x+iy = g(i)$,

$\theta = \arg(\gamma i + \delta)$, and $h = \begin{bmatrix} y^{1/2} & xy^{-1/2} \\ 0 & y^{-1/2} \end{bmatrix} \begin{bmatrix} \cos\theta & -\sin\theta \\ \sin\theta & \cos\theta \end{bmatrix}$. Prove that $h=g$ by showing that $h(i)=g(i)$ and $Dh(\zeta_0)=Dg(\zeta_0)$, where ζ_0 is the unit tangent vector to the imaginary axis at the point i.

2.4. Let $f(z) = \frac{az+b}{cz+d} \in PSL(2,\mathbb{R})$. Choose local coordinates in $S\mathcal{H}$ to be $(z,\zeta)=(x,y,\xi,\eta)$, and $(f(z), (Df)(\zeta)) = (u,v,\xi',\eta')$. Then the Jacobian matrix for the change of variables is

$$\begin{bmatrix} \frac{\partial u}{\partial x} & \frac{\partial u}{\partial y} & 0 & 0 \\ \frac{\partial v}{\partial x} & \frac{\partial v}{\partial y} & 0 & 0 \\ 0 & 0 & \frac{1}{(cz+d)^2} & 0 \\ 0 & 0 & 0 & \frac{1}{(cz+d)^2} \end{bmatrix} =$$

$$\begin{bmatrix} \frac{\partial u}{\partial x} & \frac{\partial u}{\partial y} & 0 & 0 \\ -\frac{\partial u}{\partial y} & \frac{\partial u}{\partial x} & 0 & 0 \\ 0 & 0 & \frac{1}{(cz+d)^2} & 0 \\ 0 & 0 & 0 & \frac{1}{(cz+d)^2} \end{bmatrix}$$

by the Cauchy-Riemann equations. The metric on $S\mathcal{H}$ given at the end of §2.1 is a norm in the tangent space to $S\mathcal{H}$:

$$\|(dx,dy,d\xi,d\eta)\|^2 = \frac{(dx)^2+(dy)^2}{y^2} + (d\theta)^2.$$

The invariance of this norm follows from the fact that the determinant of the upper-left block of the Jacobian matrix is equal to $|f'(z)|^2 = |cz+d|^{-4}$ and from the linearity in the last two variables.

2.8. Use Theorem 2.2.6 and Corollary 2.2.7.

2.10. Use the fact that the ring of integers in $\mathbf{Q}(\sqrt{2})$ is a Euclidean ring to construct a sequence of elements in Γ which converges to the identity.

2.11. If Γ is elementary, for any $T,S\in\Gamma$, $<T,S>$ is elementary. Conversely, suppose Γ is not elementary. By Theorem 2.4.4 it contains a hyperbolic element T with fixed points α and β. Since Γ is not elementary, there exists $S\in\Gamma$ which does not leave the set $\{\alpha, \beta\}$ invariant. Hence $<T,S>$ is not elementary.

2.13. First construct two elements in Γ: a hyperbolic element T with fixed points $\{\alpha, \beta\}$, and an element S which does not leave the set $\{\alpha, \beta\}$ fixed as in 2.10 above. Suppose first that the sets $\{\alpha, \beta\}$ and $\{S(\alpha), S(\beta)\}$ do not intersect. In this case, the elements T and $T_1=STS^{-1}$ both are hyperbolic and have no common fixed point. The sequence $\{T^nT_1T^{-n} \mid n\in\mathbb{Z}\}$ consists of hyperbolic elements with fixed points $T^nS(\alpha)$ and $T^nS(\beta)$ which are pairwise different. If the sets $\{\alpha, \beta\}$ and $\{S(\alpha), S(\beta)\}$ have one point of intersection, say α, then $P=[T,T_1]$ is parabolic (show!) and it fixes α. Since $\{\alpha\}$ cannot be Γ-invariant there exists $U\in\Gamma$ not fixing α; then $Q=UPU^{-1}$ is parabolic and does not fix α. Therefore Q and T have no common fixed points. Then for large n, elements T and Q^nTQ^{-n} are hyperbolic and have no common fixed point, and the problem is reduced to the first case.

2.14. (i) $\Rightarrow$ (ii) by Corollary 2.2.8. (i) $\Rightarrow$ (iii) by the definition.
(i) $\Rightarrow$ (iv). Suppose (iv) does not hold. Then Γ contains an element T of infinite order. If x is its fixed point then the sequence $\{T^n(z) \mid n\in\mathbb{Z}\}$, is dense on the hyperbolic circle of radius $\rho(x,z)$ centered at x. Since Γ is not elementary, there exists $S\in\Gamma$ such that $S(x)\neq x$; then the sequence $T^nS(x)$, which consists of elliptic fixed points of the following sequence $(T^nS)T(T^nS)^{-1}$, has limit points in $\mathcal{H}$, a contradiction.
(iii) $\Rightarrow$ (iv). Suppose (iv) does not hold. Then after corresponding conjugation our group must contain an element of the form

$T(z)=\exp(2\pi i\theta)z$, where θ is irrational. Since the numbers $\exp(2\pi i\theta)$ are dense on the unit circle, for an appropriate subsequence, $T^n \to \mathrm{Id}$.
(iv) $\Rightarrow$ (i). Let us consider Γ as a group of matrices, and suppose Γ_0 is a finitely generated subgroup of Γ. By a result of Selberg [Se], Γ_0 contains a subgroup Γ_1 of finite index which contains no elliptic elements of finite order. Hence by our hypothesis, it contains no elliptic elements. Thus by Theorem 2.4.5, Γ_1 is discrete. Since Γ_1 is of finite index in Γ_0, Γ_0 is also discrete; and by Theorem 2.4.8, Γ is discrete.

3.1. Use Theorem 1.2.8(ii) for the hyperbolic distance.

3.3. Let T be of the form (3.3.2). Then I(T) is given by $|z+\frac{\bar{a}}{\bar{c}}|=\frac{1}{|c|}$. It is orthogonal to the principal circle since $|a|^2-|c|^2=1$.

3.4. In general, isometric circles do not map to isometric circles under conjugation, for some exceptions see Exercise 3.6, below.

3.6. Notice that conjugation by a translation $z\to z+\beta$ maps an isometric circle to an isometric circle of the same radius translated by $-\beta$. Hence, without loss of generality, we may assume that the isometric circles for T and T^{-1} are symmetric with respect to the imaginary axis. Use formula (3.3.5) to complete the proof.

3.7. Use formula (3.3.5) again.

3.8. This follows from Theorem 2.4.3 which gives a complete description of all elementary Fuchsian groups.

3.9. Suppose F is locally finite. Then each point $z\in\mathcal{H}$ has a compact neighborhood V such that (ii) and (iii) hold. Decreasing V, if necessary, we may assume that for all $1 \leq i \leq n$, $z \in T_i(F)$, i.e. (i) holds. Conversely, if K is a compact subset of $\mathcal{H}$, for each $z \in K$ choose a V_z satisfying (i) - (iii), and choose a finite subcover $\bigcup_i V_{z_i} \supset K$.

3.10.If x lies on the side of the Dirichlet region $F = D_p(\Gamma)$, then there exists $T_x \in \Gamma$ such that $\rho(p, x) = \rho(T_x p, x)$, hence $\rho(p, x) = \rho(p, T_x^{-1} x)$. Now assume that a vertex $v \in \mathcal{H}$ is not isolated, i.e. there is a sequence of vertices $v_i \in F$ such that $v_i \to v$. According to the above remark, choose T_i such that $\rho(p, v_i) = \rho(p, T_i v_i)$. We have
$\rho(v, T_i v_i) \leq \rho(v, v_i) + \rho(v_i, T_i v_i) \leq \rho(v, v_i) + \rho(v_i, p) + \rho(p, T_i v_i) =$
$\rho(v, v_i) + 2\rho(v_i, p) \leq \rho(v, v_i) + 2\rho(v_i, v) + 2\rho(v, p)$. Hence for any $\epsilon > 0$ $\rho(v, T_i v_i) < 2\rho(v, p) + \epsilon$, for large i, which means that $T_i v_i \in K$ for all $i > N$ where K is a compact region in $\mathcal{H}$. So there is a subsequence $T_n v_n \to b \in \mathcal{H}$, but this contradicts the local finiteness of F.

3.11.Since F is locally finite, apply Exercise 3.9. Notice that if $z \in \partial F$, then $z \in T_1(F) \cap \ldots \cap T_n(F)$ for $n \geq 2$.

3.12.Let $T = \begin{bmatrix} a & b \\ c & d \end{bmatrix} \in SL(2, \mathbb{Z})$ be of order 2. Then $a + d = 0$, and the only fixed point in $\mathcal{H}$ is $z = -\frac{d}{c} + \frac{i}{|c|}$. Since $\mathrm{Im}(z) \geq \frac{\sqrt{3}}{2}$ and c is an integer, we obtain $c = \pm 1$. But then since $|\mathrm{Re}(z)| \leq \frac{1}{2}$, we obtain $d = 0$, and $z = i$.

3.13.Suppose $F \cap T(F)$ contains 3 points not belonging to the same geodesic. Since F is convex, it follows that $F \cap T(F)$ contains a hyperbolic triangle of non-zero area; hence property (ii) of the definition is violated. Hence $F \cap T(F)$ is a geodesic segment containing $T(s)$, and hence $F \cap T(F) = T(s)$.

4.2. Use Exercise 1.1 to map $z(t)$ into the imaginary axis.

4.3. This becomes obvious if we map the axis of T into the imaginary axis using Exercise 1.1: $\omega(b)$ will be mapped to a Euclidean circle tangent to the real axis, and T will become $z \to \lambda z$.

4.4. Assume that $\xi=\infty$, and its stabilizer is generated by T_0: $z \to z+1$. Let $K_0=\{x+iy \mid y\geq 2\}$, $K=\{x+iy \mid y>1\}$, and $K_1=\{x+iy \mid 1\leq y\leq 2\}$. Since Γ is non-elementary, by Excersice 2.11 and Theorem 2.4.8, for any $f\in\Gamma$, $f(z)=\frac{az+b}{cz+d}$, $<T_0, f>$ is non-elementary and discrete. Then the Jørgensen inequality (Theorem 2.4.6) holds for T_0 and any $f\in\Gamma$, and by Exercise 2.12, $|c|\geq 1$. Then $\mathrm{Im}(f(i))\leq 1$, hence $\Gamma i \cap K_0=\emptyset$, and therefore $\bigcup_{f\in\Gamma} f(K_0)\neq \mathcal{H}$. It follows that if $f(\widetilde{F})\cap K\neq\emptyset$, then $f(\widetilde{F})\cap K_1\neq\emptyset$. Suppose $\widetilde{F}$ intersects $h(K)$, then $h^{-1}(\widetilde{F})$ intersects K, and hence K_1. Let $E=\{x+iy \mid 0\leq x\leq 1, 1\leq y\leq 2\}$, then $\bigcup_n T_0{}^n(E)=K_1$, and for some n $T_0{}^n h^{-1}(\widetilde{F})$ intersects E. Since E is compact and F is locally finite, only finitely many images of $\widetilde{F}$ intersect E, say $g_1(\widetilde{F})$, ... , $g_n(\widetilde{F})$. This means that $T_0{}^n h^{-1}=g_j$ for some j, and $h(K)=g_j{}^{-1}(K)$. Now suppose z_1, z_2, ... is a sequence of points in K such that $\mathrm{Im}(z_n) \to \infty$. Then there are elements h_1, h_2, ... $\in \Gamma$ such that $h_n(z_n)\in\widetilde{F}$, i.e. $\widetilde{F}$ intersects each $h_n(K)$. By the previous argument, $h_m(K)=g_{j_m}{}^{-1}(K)$, i.e. $h_m(K)$ may take only finitely many values, hence (by going to a subsequence and renumbering) we have $h_1(K)=h_2(K)=$ ... . It follows that there are integers t_2, t_3, ... such that $h_m=h_1T_0{}^{t_m}$. Thus $h_1(w_m)\in\widetilde{F}$, where $w_m=T_0{}^{t_m}(z_m)$, and since $\mathrm{Im}(w_m)=\mathrm{Im}(z_m) \to \infty$ we conclude that $w_m \to \infty$. Hence $h_1(\infty)\in\widetilde{F}$.

4.6. Recall that Π is a star-like polygon: there exists $p\in\Pi$ such that geodesic rays from p to the vertices lie inside Π. Use these rays to divide Π into n triangles and apply Theorem 1.4.2.

4.7. Use the unit disc model, and let g_1 and g_2 be two geodesic rays passing through 0 making the angle α. For any point $P \in g_1$ there exists a unique geodesic $g_3(P)$ through P such that the angle at P is equal to β. If P is sufficiently close to 0, $g_3(P)$ intersects g_2 (why?) at an angle $\gamma(P)$. Use continuity arguments to show that for a unique $P \in g_1$, $\gamma(P)=\gamma$.

4.9. If $g \geq 2$, then $\mu(F) \geq 4\pi$. If $g=1$, then as $\mu(F)>0$, F must have periods and hence $\mu(F) \geq \pi$, the minimum is attained for a group with signature (1; 2). If $g=0$, then

$$\mu(F)=2\pi\Big(-2+\sum_{i=1}^{r}\Big(1-\frac{1}{m_i}\Big)\Big).$$

Since $1-\frac{1}{m_i} \geq \frac{1}{2}$, $\mu(F) \geq 2\pi(-2+\frac{r}{2})=\pi(r-4)$ so that if $r \geq 5$, then $\mu(F) \geq \pi$, the minimum attained for a group with signature (0; 2, 2, 2, 2, 2). If $r=4$ and $m_1=m_2=m_3=m_4=2$, then $\mu(F)=0$ which is nonadmissible. The minimum positive value of $\mu(F)$ for $r=4$ corresponds to the signature (0; 2, 2, 2, 3) giving $\mu(F) \geq \pi/3$. Let us consider the remaining case $g=0$, $r \leq 3$. If $r \leq 2$, then $\mu(F)<0$. Hence the only case to consider is $g=0$, $r=3$. In this case Γ is a triangle group with signature $(0; m_1, m_2, m_3)$, and

$$\mu(F)=2\pi\Big(1-\frac{1}{m_1}-\frac{1}{m_2}-\frac{1}{m_3}\Big).$$

We may assume that $m_1 \leq m_2 \leq m_3$. If $m_1 \geq 4$, then $\mu(F) \geq \pi/2$. If $m_1=3$ then a group with signature (0; 3, 3, 3) gives $\mu(F)=0$; and the positive minimum is attained for (0; 3, 3, 4) giving $\mu(F) \geq \pi/6$. If $m_1=2$, then $m_2>2$ and if $m_2 \geq 4$, then $\mu(F) \geq \pi/10$, the minimum corresponding to (0; 2, 4, 5). If $m_2=3$, then (0; 2, 3, 6) gives $\mu(F)=0$; and the positive minimum is attained for (0; 2, 3, 7), giving

$\mu(F) \geq \pi/21$.

4.12. A hyperbolic reflection is an orientation-reversing isometry of $\mathcal{H}$, and hence it is given by

$$T(z) = \frac{a\bar{z}+b}{c\bar{z}+d},$$

with $ad-bc=1$ (see Theorem 1.3.1). We know that the fixed-point set for T is the given geodesic Q. Solving $T(z)=z$, we obtain $c|z|^2+dz=b+a\bar{z}$, and equating the imaginary parts we get $d=-a$. Show that the fixed-point set is given by the equation $|z-\frac{a}{c}|=\frac{1}{|c|}$ if $c\neq 0$, and $x=-\frac{b}{2a}$ if $c=0$, a geodesic in either case. Express a,b,c, and d in terms of the center and the radius of Q to obtain the formula for inversion (3.3.5).

5.2. It is useful to introduce the *Legendre symbol* (see e.g. [S]):

$$\left(\frac{b}{p}\right) = \begin{cases} 1, & \text{if b is a quadratic resedue mod p} \\ -1, & \text{if b is a quadratic non-residue mod p.} \end{cases}$$

By the Quadratic Reciprocity Law, $\left(\frac{-1}{p}\right)=(-1)^{(p-1)/2}$. Then in our case $\left(\frac{-1}{p}\right)=-1$, hence -1 is a non-residue mod p, and this follows from Theorem 5.2.5.

5.3. If $a>0$ then $a\in(\mathbb{R}^*)^2$, and then by Theorem 5.2.1 $\left(\frac{a,\ b}{\mathbb{R}}\right)\approx M(2, \mathbb{R})$. If $a<0$, $b<0$ then $\mathrm{Nrd}(x)$ $=x_0{}^2-ax_1{}^2-bx_2{}^2+abx_3{}^2>0$ unless $x=0$, hence (by Theorem 5.2.3) in this case $\left(\frac{a,\ b}{\mathbb{R}}\right)$ is a division algebra.

5.5. Use also Exercise 5.3.

5.6. First prove that $\Gamma^{(2)}$ is a normal subgroup of Γ; then notice that for any $T\in\Gamma/\Gamma^{(2)}$, $T^2=\mathrm{Id}$ which implies that $\Gamma/\Gamma^{(2)}$ is abelian. Since

Γ is finitely generated, so is $\Gamma/\Gamma^{(2)}$. Hence $\Gamma/\Gamma^{(2)}$ is finite. Now apply the Classification Theorem for finite abelian groups.

5.7. A basis of $A(\Gamma)$ constructed in Proposition 5.3.6 can be chosen in the form $\{1_2, \alpha^2, \beta^2, \alpha^2\beta^2\}$ where α and β are hyperbolic elements in Γ. See [T1].

5.10. Let $(c, d)=s$, $c=ks$. Check that $q=\frac{k}{(k,s)}$ works.

5.11. Show that $K_n/K(n)$ is isomorphic to the group of all diagonal unimodular matrices

$$\begin{bmatrix} [a] & 0 \\ 0 & [d] \end{bmatrix},$$

where $[a]$ and $[d]$ are invertible elements in the ring of integers mod n.

5.12. Use the natural homomorphism ψ_n: $SL(2,\mathbb{Z}) \to SL(2,\mathbb{Z}_n)$ as in (5.5.1) to find $[K_0(n) : K(n)]$. Since $\psi_n(K_0(n)) = \left\{\begin{bmatrix} [a] & 0 \\ [c] & [d] \end{bmatrix} \in SL(2,\mathbb{Z}_n)\right\}$ conclude that $[K_0(n) : K(n)] = n\varphi(n)$. Now use Exercise 5.11 and (5.5.2).

5.13. Use definition (II) of an order. $\mathcal{O}$ is a free $\mathbb{Z}$-module of rank 4. In order to check that $\mathcal{O}$ is a subring, it is sufficient to check that the products of its generators belong to $\mathcal{O}$. For example, $\frac{1+j}{2}\cdot i = \frac{i-k}{2} = i - \frac{i+k}{2} \in \mathcal{O}$.

5.14. [T3] For any $T = \begin{bmatrix} a & b \\ c & d \end{bmatrix} \in \Gamma$ we have $T^2 - tT + 1_2 = 0$, where $t = tr(T) = a+d$. Hence for any integer n, $T^n = f_n(t)T + g_n(t)1_2$, where

$f_n(x)$ and $g_n(x)$ are monic polynomials in $\mathbb{Z}[x]$. Moreover, for any A, B, $T \in \Gamma$ we have

$$\begin{cases} tr(A)tr(B)=tr(AB)+tr(AB^{-1}), \\ tr(ABAT)=tr(AB)tr(AT)-tr(BT^{-1}). \end{cases}$$

For any $T \in \Gamma$, express $T=S_{i_1}^{n_1} \ldots S_{i_s}^{n_s}$, and let m(T) be the minimum of $\sum_{j=1}^{s} |n_j|$ for all such expressions. Now use induction on m(T) and the above formulae.

5.15. Use Exercise 5.14.

5.17. Use Exercises 5.16 and 5.7.

BIBLIOGRAPHY

[A] L. Ahlfors, "Complex Analysis", 2nd. ed., Mc Graw-Hill Book Co., 1966

[B] A. Beardon, "The Geometry of Discrete Groups", Springer-Verlag, 1983

[BH] A. Borel and Harish-Chandra, Arithmetic subgroups of algebraic groups, Ann. of Math. 75 (1962) 485-535

[F] L. Ford, "Automorphic Functions", Chelsea, 1951

[GP] I. M. Gelfand, M. I. Graev, I. I. Pyatetskii-Shapiro, "Representation Theory and Automorphic Functions" (English translation). W. B. Saunders, Philadelphia, 1969

[JS] G. A. Jones, D. Singerman, "Complex Functions", Cambridge University Press, 1987

[K] S. Katok, Reduction theory for Fuchsian groups, Math. Ann. Math. 273 (1986) 461-470

[L] J. Lehner, "Discontinuous Groups and Automorphic Functions", Math. Survey & A M S, Providence 1964

[M] W. Magnus, "Noneuclidean Tesselations and Their Groups", Academic Press, 1974

[Ma] B. Maskit, On Poincaré's Theorem for fundamental polygons, Adv. Math. 7 (1971) 219-230

[P] H. Poincaré, Théorie des groupes Fuchsiens, Acta Math. 1 (1882) 1-62

[S] H. Shapiro, "Introduction to the Theory of Numbers", J. Wiley & Sons, 1983

[Se] A. Selberg, On discontinuous groups in higher-dimensional spaces,–In: Contributions to Function Theory, Tata Institute, Bombay, 1960

[Sh] G. Shimura, "Introduction to the Arithmetic Theory of Automorphic Functions", Princeton University Press, Princeton 1971

[Sp] G. Springer, "Introduction to Riemann Surfaces", 2nd ed., Chelsea Publ. Co., New York, 1981

[T1] K. Takeuchi, A characterization of arithmetic Fuchsian groups, J. Math. Soc. Japan, 27 , No.4, (1975) 600-612

[T2] K. Takeuchi, On some discrete subgroups of $SL_2(\mathbb{R})$, J. Fac. Sci. Univ. Tokio, Sec. I, 16, (1969) 97-100

[T3] K. Takeuchi, Arithmetic triangle groups, J. Math. Soc. Japan, 29, No. 1, (1977) 91-106

[Te] A. Terras, "Harmonic Analysis on Symmetric Spaces and Applications", Springer-Verlag, 1985

[Ti] J. Tits, Classification of algebraic semisimple groups, Proc. Symp. Pure Math. v.9, Providence 1966, 33-62

[Vi] M.-F. Vignéras, "Arithmetique des Algebres de Quaternions", S L N 800, 1980

[VS] Э. Б. Винберг, О. Б. Шварцман, "Дискретные Группы Движений Пространств Постоянной Кривизны", Итоги Науки и Техники, том 29, Москва, 1988

[W] A. Weil, Algebras with involutions and classical groups, J. Indian Math. Soc. 24 (1960) 589-623

INDEX